JEROME C.
HUNSAKER

AND THE  RISE OF

AMERICAN
AERONAUTICS

SMITHSONIAN HISTORY OF AVIATION AND SPACEFLIGHT SERIES

Dominick A. Pisano and Allan A. Needell, Series Editors

Since the Wright brothers' first flight, air and space technologies have been central in creating the modern world. Aviation and spaceflight have transformed our lives—our conceptions of time and distance, our daily routines, and the conduct of exploration, business, and war. The Smithsonian History of Aviation and Spaceflight Series publishes substantive works that further our understanding of these transformations in their social, cultural, political, and military contexts.

JEROME C. HUNSAKER

AND THE RISE OF

AMERICAN AERONAUTICS

WILLIAM F. TRIMBLE

SMITHSONIAN INSTITUTION PRESS
WASHINGTON LONDON

Copy editor: Diane Hammond
Production editor: Robert A. Poarch
Designer: Jody Billert

Library of Congress Cataloging-in-Publication Data
Trimble, William F., 1947–
    Jerome C. Hunsaker and the rise of American aeronautics / William F. Trimble.
    p. cm. — (Smithsonian history of aviation and spaceflight series)
    Includes bibliographical references and index.
    ISBN 1-58834-006-6 (alk. paper)
    1. Hunsaker, Jerome C. (Jerome Clarke), 1886–1984. 2. Aeronautical engineers—
United States—Biography. I. Title. II. Series.
TL540.H85 T75 2002
629.13′0092—dc21
[B]   2001042038

British Library Cataloguing-in-Publication Data is available
Manufactured in the United States of America
09 08 07 06 05 04 03 02   5 4 3 2 1

♾ The paper used in this publication meets the minimum requirements of the American National Standard for Information Sciences—Permanence of Paper for Printed
Library Materials ANSI Z39.48-1984.

To Mary Jane Duncan Trimble

CONTENTS

ACKNOWLEDGMENTS

More years ago than I care to think about, Richard K. Smith, one of the most respected historians of aeronautics in the country, warned me that a biography of Jerome C. Hunsaker would be the biggest research and writing job I was likely to undertake. Naively, I replied that, yes, I knew it would be a major undertaking but that I was used to such things and that I could pull it off expeditiously.

As usual, Dick Smith was right. The book you hold here turned out to be far more challenging than I had imagined, taking me through more than a hundred boxes of archival material and countless pages of books and articles and involving more rethinking, reorganization, and rewriting than anything I had done before. The problem was the subject, a man whose career cut a wide swath through aviation history and whose hand seemed to touch every aspect of twentieth-century aeronautics. Fortunately, Dick was there to read and critique the manuscript, forcing me to sharpen my thinking and my prose at a crucial early stage. This book, and earlier ones I have written, owe much to Dick's bottomless intellect and common sense.

At the Smithsonian's National Air and Space Museum, Thomas Wildenberg, a skilled historian and author and a former Ramsey Fellow in Naval Aviation History, read and critiqued the manuscript, his probing questions helping me place Hunsaker's naval career in perspective. Dominick Pisano, Peter L. Jakab, John D. Anderson Jr., Alex Spencer, Von Hardesty, Allan A. Needell, William Althoff, and Tom D. Crouch, at the National Air and Space Museum, offered advice and support at key stages. The museum's Michael J. Neufeld arranged for a Smithsonian short-term visitor appointment at an early stage of the project and smoothed the way for me to take the Charles A. Lindbergh Chair of Aerospace History in 1999–2000. At the museum's archives, Marilyn Graskowiak and Larry Wilson made the extensive Hunsaker papers available before the collection was fully processed and cataloged.

At other repositories archivists led me through a sometimes daunting maze of Hunsaker materials. Elizabeth Andrews, at MIT's Institute Archives and Special Collections, fielded my questions early in the project and promptly dispatched photocopies from the Hunsaker papers. She and others in Cambridge cheerfully met all my demands during a week-long research trip in the spring of 1997. Aaron Vidaver tracked down municipal records in Boston pertaining to the Hunsaker home in Louisburg Square. Barry Zerby, at NARA's Archives II in College Park, Maryland, patiently accommodated all of my requests, no matter how ill timed or ill conceived they might have been. His knowledge of the Bureau of Aeronautics records at the National Archives is unsurpassed. Marjorie Ciarlante, at Archives II, helped me with NASA records. At the "downtown" Archives I, Richard W. Peuser and Rebecca Livingston identified and made available Hunsaker personnel records that I otherwise would have missed. Roger D. Launius, chief historian at NASA, made Hunsaker and NACA materials available to me at his office in Washington. During a trip to the University of Akron Archives, I was led by John V. Miller to the Hunsaker materials in the voluminous Lighter-than-Air Society Collection. Beverly Lyall, at the U.S. Naval Academy Archives, supplied photocopies documenting Hunsaker's academy career.

Many others made this book possible. Mark Levinson's insightful critique of the aeronautical engineering and MIT chapters early in the project was of immense help, as was the assistance of the engineer and historian Earl A. Thornton. Mark Gatlin, the aviation and military history editor at the Smithsonian Institution Press, provided encouragement and direction while I labored through revisions of the manuscript. Diane Hammond skillfully copyedited the final product. I also want to express deep thanks to the following for sharing their knowledge and advice: Harold ("Hal") Andrews, Eric Feron, Thomas C. Hone, Jerome C. Hunsaker Jr., Jerome C. Hunsaker III, Charles L. Keller, Jeremy R. Kinney, Michael McCormick, Lee Pearson, J. Lawrence Lee, David R. Mets, Henry Cord Meyer, Anne Millbrooke, Leonard S. Reich, Eric Schatzberg, David Bryan Taylor, David Vogelgesang, and Jacob Vander Meulen. If I have omitted anyone, it is wholly unintentional.

I will forever hold in my memory a delightful luncheon and visit, on a cold December day in 1996, with Alice Hunsaker Bird and her family at their charming seventeenth-century home in the hills above Henley-on-Thames. The family's warm welcome was in happy contrast to the weather. Alice, who has written a superb Hunsaker family history, which includes a collection of documents about her father's life, answered my questions about her father readily and with good humor.

Grants came at key junctures in my work, starting with research money from the Air War College in the spring of 1994 and ending with additional funding in the spring of 1999. My thanks go to Bryant P. Shaw and Edward G. Holland, commanding officers, department heads, and colleagues. Auburn University provided a research grant-in-aid, allowing me to visit MIT in the spring of 1997. In the Department of History, chair James R. Hansen, a first-rate aerospace historian in his own right, supplemented the university's funding and juggled departmental teaching schedules to provide me with time for research and writing.

Paul and Harriet Sirovatka and their son, Jonathan, opened their Kensington, Maryland, home to me on my research trips to Washington. Friends for decades, the Sirovatkas have been a generous and supportive second family during this and previous book projects. Words cannot express how much I owe them.

My first family somehow endured this book, as they have others. To my wife, Sharon, and sons, Will and Mike, go my sincerest thanks for their patience and forbearance during my many absences from home and long hours at the computer keyboard. This book would not have been possible without their love and support.

1.   An Officer and an Engineer

I n the 1970s, more than twenty years after his retirement from the Massachusetts Institute of Technology, Jerome C. Hunsaker continued to spend
time in his office on the second floor of the four-story, brick and limestone
Guggenheim Building on the MIT campus. Declining health had forced him
to abandon his daily walks from his home on Louisburg Square, in the genteel
Beacon Hill section of Boston, to Cambridge, on the other side of the Charles
River. Now, in his late eighties, his age and infirmities meant that he usually had
to be driven. Once settled in his sparely appointed office he puffed on his pipe
while he read and answered correspondence, spoke to friends and colleagues
on the telephone, and chatted with students, few of whom had any inkling of
the role he had played in the development of aeronautics.

A series of strokes had stolen Hunsaker's short-term memory to the degree
that he often had trouble recognizing even close friends and colleagues, yet the
old man could recall with remarkable clarity events and people from his six-
decade professional career. When Walter T. Bonney interviewed him in November 1971 for a history of the National Advisory Committee for Aeronautics
(NACA), Hunsaker pointed to a photograph of Admiral David W. Taylor and
recalled how, in 1916, when he was a young naval officer completing his
doctoral degree in aeronautical engineering at MIT, Taylor had called him to
Washington to take responsibility for all of the navy's aircraft design and procurement. Taylor, Hunsaker said, had single-handedly brought the navy into the
modern technical world.[1]

Jerome Hunsaker came to maturity not long after the turn of the century and barely a generation after engineering was transformed from a craft into a profession. He viewed himself as a member of a distinct technical elite and his work as a means of bringing about social and economic progress. His ideology proposed that engineering methodology could solve any number of social and economic problems.[2] That said, his and other engineers' rationality was circumscribed in ways that bred conservatism and resistance to change. Hunsaker's fascination with the rigid airship in the 1930s and his failure to accept the political realities of space and rocket technology in the 1950s stand in paradoxical contrast to his progressive worldview.

Hunsaker strode confidently into the exciting new "high-tech" world of aeronautics. His unwavering faith in a systematic, engineering approach to problems in aeronautics meshed with contemporary developments in Europe, where in the years before World War I British, French, and German experimenters reconstituted aerodynamic research as "engineering science." Mark Levinson defines engineering science as "science distinctly concerned with the problems of engineering, often based on existing science . . . and making use of clearly enunciated rational approximations when necessary."[3] No one in the United States did more than Hunsaker to foster engineering science, transferring European theories of aerodynamics to America and bridging the gap between theory and practice. He once wrote, "The aeronautical engineer practices his art under the constant and often passionate criticism of the operator, and faces final evaluation by the dispassionate but often cruel laws of nature."[4]

Throughout Hunsaker's life, the rationality and discipline inherent in engineering formed the bedrock of a career that took him into the military bureaucracy, private industry, academic administration, and public policy. Fresh from MIT, he went to Washington to bring order to the navy's World War I aircraft design and procurement programs and was a catalyst in the administrative reform of naval aviation in the tumultuous years after the war. After leaving the navy, Hunsaker spent forty years in a career that placed him squarely within what became known as the military-industrial-university complex.

At Bell Labs in the late 1920s, Hunsaker helped fashion a modern commercial airway network; during his years with Goodyear-Zeppelin he proposed a rational vision of transoceanic air travel, albeit with huge rigid airships rather than multiengine, long-range airplanes. Under his guidance MIT in the 1930s and 1940s took advantage of private and public largesse to build a world-class program of aeronautical engineering rooted in basic research. Last, and hardly least, in his multifaceted career in aeronautics was his chairmanship of the NACA from 1941 to 1956. He viewed this post as an opportunity to apply on a national level his philosophy about the rational progression of aeronautics and basic aeronautical research. During those years, he directed the NACA's resources to fighting the second of the century's global conflicts before turning his formidable intellect and political know-how to building it into one of the world's premier research organizations.

Hunsaker was the archetypal heterogeneous engineer. Definitions vary, but it is easiest to think of a heterogeneous engineer as comfortable not only in his or her technical field but also in the wider world and able to understand and deal with technology's social, cultural, political, and economic ramifications. Hunsaker's understanding of the personal, professional, and institutional advantages of forging close relationships outside the confines of the research laboratory defined much of his career in the world of business and government.[5] For better or worse, Hunsaker and others like him raised the engineer to prominence in a world in which technology was woven into the social, economic, and political fabric, a world defined by the scientifically provable, the technically possible, and the politically feasible. His long, rich life and successful work in aeronautics centered on those verities and on little else. A survey of Hunsaker's life and career opens a window, however imperfectly, on a man who found in aeronautics not only personal and professional satisfaction but also enduring proof of the promise of technology in the twentieth century.

Jerome Hunsaker's roots lay in the Midwest, the heartland of a nation undergoing profound changes in the years after the Civil War. In the 1870s, Creston, Iowa, in the rolling hills of Union County, in the southwestern part of the state, was a town of about five thousand people. As a division point on the main line of the Chicago, Burlington, and Quincy Railroad, the town attracted a generation of settlers from the East and from other parts of the Midwest, drawn by the railroad's promise of work (and also by the lure of cheap, rich farmland). One of the newcomers was Walter J. Hunsaker, who arrived in 1879 from Carthage, Illinois, to take over publication of the *Creston Republic*.[6]

Generations earlier, Walter's direct ancestor, Hartmann Hunsaker, had been forced to leave Switzerland because of his Anabaptist beliefs. These early Hunsakers lived for a time in the Palatinate and in Holland before sailing for Philadelphia in 1731. From Philadelphia they moved to Lancaster County, where Hartmann died a few years later. His son, Johannes, settled in western Pennsylvania in 1784 and then, after the death of his wife, in Kentucky. Nicholas, one of Johannes's sons, fathered eight children, one of whom, Jacob, moved back east to Fayette County, Pennsylvania. There, at the foot of Laurel Mountain, in 1820, George Troutman Hunsaker, Walter's father and Jerome's grandfather, was born. George worked for a time as a tinsmith in Cincinnati; he then spent several years on Ohio and Mississippi river steamboats before returning to Cincinnati and marrying Emeline Coddington. Not many years later, he was off again with his family to Kentucky, Missouri, and Iowa. In 1857, while the

Hunsaker clan was in Keokuk, Iowa, a son, Walter Jerome, was born. The Hunsakers did not stay long in Keokuk, moving in 1860 across the Mississippi and inland to Carthage, Illinois, where George finally settled down and opened a hardware store.[7]

The newspaper bug infected Walter Hunsaker while he was attending Carthage College, and from 1872 until the end of his life he could not remember a day coming home without ink under his fingernails. Becoming publisher of the *Creston Republic* was an important step, providing Walter income, independence, and experience in the profession that occupied the rest of his life. In Creston, he met Alma Lyle Clarke, and on 21 October 1885 they were married. Born in 1863, Alma was the oldest of six children of Henry Sayward and Docela Weaver Clarke. Henry Clarke was a Chicago grain broker who had moved to Creston following the Civil War, invested in a bank, and earned modest wealth and considerable status in the community. Shortly after his marriage, Walter decided to take a job as night editor of the *Detroit Tribune,* which meant relocating to a strange new city. When Alma became pregnant, she and Walter agreed that she should return to Creston, where their parents still lived, to have the baby. Jerome Clarke Hunsaker was born at the Clarke home in Creston on 26 August 1886.[8]

Jerome was the Hunsakers' only child. He was reared in an atmosphere of doting toleration tempered by exasperation with his precociousness. In Detroit, the Hunsakers lived in a rented house on Third Avenue, along which horse-drawn streetcars ran. Strategically located between Lake Huron and Lake Erie, Detroit at the time was a quietly prosperous small city dependent on commerce and light manufacturing. Hunsaker remembered Farquhar's grocery store across the street as "a fascinating place with barrels of sugar, flour, vinegar . . . and cheese in a great wheel, from which portions were cut." He attended Irving School in the neighborhood and later bicycled to Central High School.

Among the high school courses Jerome found interesting were physical geography and American history; science and math were not his strengths. He was not especially close to his father, although he shared his interest in ornithology and became a skilled taxidermist, mounting birds that his father brought back from hunting expeditions. An agnostic, Walter impressed on his son an abiding skepticism about religion and a determination to look to reason for answers to the mysteries of the universe. Hunsaker's daughter Alice recalls that later in his life her father attended Christmas and Easter services out of a sense of family obligation but, otherwise, had "absolutely no interest" in religion of any sort.[9]

Alma and little Jerome spent the summers in Creston, usually leaving Walter behind in Detroit. Hunsaker remembered that he did not get along well with

Jerome C. Hunsaker's father, Walter J. Hunsaker. In the 1890s Walter was a prominent newspaper editor in Detroit. (Courtesy JCH/MIT)

his cousins and that he was a closer friend of the neighborhood boys, one of whom had a talking crow. Later, the Clarkes acquired a cottage on Lake Okoboji, where Jerome mastered a catboat and a small racing sailboat. There, too, he discovered the joys of fishing and hunting, regularly bagging ducks, bitterns, and muskrats. He even tried trapping. Using a homemade deadfall he once trapped a skunk in his grandfather's barn in Creston, with odoriferous results; years later he recalled that the skunk episode made him for a time "very unpopular" in the Clarke household.[10] From those days to the end of his life, Hunsaker was an enthusiastic outdoorsman, passionately dedicated to the rod and gun.

In early 1902 the Hunsakers moved to Saginaw, a small city on the Saginaw River at the base of Michigan's "thumb," where Walter Hunsaker became editor and co-owner of the *Saginaw Courier-Herald*. Saginaw was then undergoing a painful transformation from a wide-open, rough-and-tumble boomtown, wholly

Jerome, aged four or five, after the Hunsakers had moved to Detroit. He appears unhappy with the then-common practice in formal photography of dressing young boys
as girls. (Courtesy Alice Hunsaker Bird)

Jerome as a youth in Detroit, by his dress seemingly ready for a career at sea. (Courtesy Alice Hunsaker Bird)

dependent on the rich timber resources of the area, into a modern industrial city. While other lumber towns withered and died, Saginaw found new life when a fresh generation of entrepreneurs turned their attention to financing plate-glass factories, iron foundries, and manufacturing enterprises specializing in furniture, barrels, windmills, and washboards.[11]

The move from Detroit to Saginaw was not an easy one for Walter. The newspaper business, then as now, was highly competitive and involved considerable capital costs and a great deal of risk. Walter Hunsaker's co-ownership of the *Courier-Herald* was the direct result of his friendship with Chase S. Osborn, who had published a paper in Sault Sainte Marie and had made a small fortune in the iron ore business before buying the Saginaw paper. Osborn stayed in the "Soo" to pursue a political career that climaxed with his election as Republican governor, leaving the day-to-day finances and operation of the paper in Walter's capable hands. Walter's connections in the Republican party included not only Osborn but also Joseph Fordney, who represented the Eighth Congressional District and who had authored numerous protective tariffs intended to shield American agriculture and industry from foreign competitors.[12]

Jerome's years in the family home on North Jefferson Avenue in Saginaw were happy ones. Fascinated by the power of the wind, he fabricated a "sailing wagon," which he briefly ran on the streets of Saginaw in 1903. The experiment led to much consternation among pedestrians and the operators of horse-drawn vehicles, and the local police finally ended his escapades with a gentle reprimand. At Saginaw High School, which had an excellent reputation in the state for its science curriculum, he joined the Lambda Sigma fraternity and, during his senior year, edited the class yearbook. Meanwhile, his grandfather Clarke sold the cottage on Lake Okoboji and bought an old hotel and farm on Rest Island, in the Mississippi River near Lake City in southeastern Minnesota. After this the Hunsakers usually spent at least part of their summers with Alma's family at Rest Island, Jerome whiling away the hours hunting, fishing, and sailing.[13]

In 1903, during his junior year in high school, Jerome became interested in the U.S. Naval Academy when one of his friends applied for and received an appointment (unfortunately, the friend failed the entrance examination). Undeterred, Jerome applied the following year, receiving his appointment to the academy from his father's friend Joe Fordney. Jerome benefited from recent legislation aimed at supplying the larger number of officers needed for the greatly expanded navy: for ten years, starting in 1903, each congressman and senator could appoint two candidates to Annapolis, instead of the one previously allowed. When Jerome learned of his appointment, he dropped out of school and left for Annapolis, failing to complete his senior year. Many years later—in 1936—his alma mater awarded him an honorary diploma.[14]

Jerome about the time of his senior year at Saginaw High School. He left for the Naval Academy before graduation. (Courtesy Alice Hunsaker Bird)

In Annapolis, the seventeen-year-old Hunsaker found a room in William Handy's boardinghouse and enrolled in a "cram" school to prepare for the academy's tough entrance examination. After only a few weeks, confident that he would have no trouble with the exam, he left school, rented a boat, and spent a month sailing around the Chesapeake Bay—an "education of a different kind," he recalled. Back in Annapolis, he took the examination on 1 July and gained admittance four days later, with the 296 other young men in the class of 1908. Like many of his classmates, Midshipman Hunsaker was filled with enthusiasm, although he had no idea what the academy or the navy were like or what he would do as a naval officer. At the time, the academy was undergoing major physical changes. Gray granite buildings, in various states of completion, now replaced the ramshackle wood and brick buildings of the academy on the Severn River. One of the most impressive of the new buildings was a massive dormitory, later named Bancroft Hall, on which work had begun in 1901. Because Bancroft was not yet ready for occupancy, Hunsaker lived in temporary quarters until September, when he moved into Bancroft with one of the two battalions composing the brigade of midshipmen.[15]

In other ways, too, Hunsaker was caught between the old and the new. For decades, the academy had been mired in academic mediocrity, but with a new century and a new globe-girdling steel navy came a rigorous curriculum that placed a premium on achievement in science and engineering. Despite the high standards, the academy emphasized rote learning rather than original thinking. As much as the physical and intellectual atmosphere at Annapolis had evolved, the culture of the academy had not substantially changed since the previous century. Incoming plebes—known as *midshipmen* after the term *naval cadet* fell out of use in 1902—entered a strange new world, marked by strict discipline, a hierarchical social structure, communal living quarters, and a moral system founded on an almost feudal sense of honor and duty. The battalion dominated the academy's social and command structure. Encompassing a cross section of midshipmen from first classmen through plebes, the battalion marched to class across the Yard and dined at the same tables in Bancroft Hall's cavernous mess. The battalion, more than anything else at the academy, inculcated a sense of belonging and group awareness. Friendships formed there that lasted through whole careers if not lifetimes.

But there was a price to be paid, too, exacted in a regimen of uniformity and a ruthless suppression of individual identity. Those who did not conform usually "bilged," or washed out—among them was Chauncey Ripley, Hunsaker's roommate during his plebe year. But those who made it through four years emerged with a sense of self-worth and tradition and a willingness to sacrifice immediate personal wants and needs for the greater good of their comrades and the service. These characteristics were just as essential for the efficient functioning of modern steam warships as they were for the square-riggers of yore. In such an atmosphere, it was not surprising that the young men accepted hazing as a means of "building character." Known at the time as "running," hazing at the academy was probably no better and no worse than it was at any college or preparatory school at the time, but because Annapolis was a national institution there was a strict injunction against the custom. Nevertheless, the practice was rampant, with plebes being the most frequent victims.[16]

Neither a grind, who retreated into his academic work, nor a "mid," whose sole focus was the social and athletic whirl, Hunsaker carved out a satisfactory middle ground. Like other midshipmen, Hunsaker acquired a nickname — "Honey," or "Hunny" — although most of his friends called him Jerry. He remembered that life at Annapolis "was pretty happy and lively," punctuated by the occasional dance, or "hop," to which he and other midshipmen "dragged" the local "Crabtown" girls. At the academy there was a wide offering of intramural sports. Five feet, nine inches tall and weighing about 145 pounds, Hunsaker tried out for and made the track team, doing "pretty well" in the mile. But even if he did not excel in sports, Hunsaker was a star in academics. At the end of his first year he stood second in his class. Among his best subjects were mathematics, English, and modern languages; he performed less well in mechanical drawing and had poor marks in efficiency and conduct. During his plebe year, he received demerits for most of the usual offenses, among them not shining his shoes, "shifting in the ranks," "skylarking," and talking during brigade formations.[17]

The third-class cruise in the summer of 1905 was almost like a reward for surviving the plebe year. Hunsaker joined the complement of the steam sloop *Hartford,* Adm. David Farragut's flagship at the Battle of Mobile Bay. The venerable warship slipped its moorings at Annapolis on 5 June, stood down the Chesapeake Bay, and sailed up the East Coast as far as Maine. Hunsaker's assignment was one of the topsail yards. He recalled "the fear and trembling with which we ran up the ratlines to furl the sail. . . . One held on with white knuckles for oneself and used the other hand for Uncle Sam." Still he was perceptive enough to understand that the menial tasks assigned to the midshipmen were valuable "training and conditioning for responsibility, which most academic education fails to provide." Before the cruise was over, Hunsaker fell ill with bronchial pneumonia and had to leave the ship. In the days before antibiotics, pneumonia could be a killer; Hunsaker was not given a clean bill of health and discharged until the end of August.[18]

The next summer, 1906, Hunsaker found himself aboard the relatively new ten-gun, 3,200-ton, protected cruiser *Cleveland.* The ship left Annapolis on 17

The USS *Hartford*, Jerome Hunsaker's ship during his third-class cruise in the summer of 1905. The old steam sloop had been Admiral David Farragut's flagship at the Battle of Mobile Bay in 1864. (19N-11431, Courtesy National Archives)

June, crossed the Atlantic to Madeira and the Azores, and then sailed back again to Bar Harbor, Maine, where Hunsaker found a batch of letters and a box of candy from home waiting for him when the ship arrived on 25 July. He saw little that was remarkable about Bar Harbor but did note that on the heights overlooking the town were "hundreds of the most pretentious summer palaces." Following some sightseeing, he and a few of his shipmates ate a "scrumptious" dinner at the Louisbourg Hotel, finding it a welcome change from the canned "grub" they normally consumed at sea. Aboard ship, Hunsaker had whetted his engineering appetite by keeping a technical diary that featured professional-looking drawings of the ship's ammunition hoists, six-inch guns, communication systems, and propulsion machinery.[19]

At the end of his second year, Hunsaker moved up to first in his class, with superior grades in mathematics, physics, chemistry, English, and modern languages. The biggest problem he had at the academy occurred midway though his junior year. On 13 January 1907 Hunsaker was on night watch with another man, who collapsed after drinking too much. Hunsaker let the man's roommate relieve him and did not disclose the incident, a clear violation of orders. He was

given fifty demerits for "deliberate neglect of duty." In a letter to the commandant in June, Hunsaker pleaded that the word *deliberate* did not fit the case, explaining that he had been ignorant of the circumstances behind the substitution
and that he had not reported the offense because he did not fully comprehend
its seriousness. Hunsaker admitted many years later that permitting the substitution was the result of a foolish "schoolboy loyalty to a classmate that conflict[ed] with the spirit of responsibility which is the basis for discipline." The
incident, however, did no irreparable harm to his otherwise exemplary record.
At the conclusion of his junior year, he ranked second in his class, with exceptional grades in physics, mechanics, navigation, ordnance, and seamanship.[20]

Hunsaker began the 1907 summer cruise as soon as classes ended in the first
week of June. This time his ship was the *Arkansas,* whose skipper was Comdr.
Bradley A. Fiske, possibly the brightest and most innovative officer in the service at the time. Fiske gave Hunsaker his first taste of advanced technology and
the new navy. At Fiske's invitation, Hunsaker experimented with an improved
optical turret range finder, following up with his own studies of the complex
mathematical problems of gunnery at sea. Fiske also sought a solution to the
problem of determining the height of an astronomical object above the sea
when weather conditions obscured the horizon. Using a small boat as a reference point for the calculations, Hunsaker correlated the results from observations of the horizon on a clear day and verified the technique as a supplement
to conventional navigation procedures. Fiske published the results of the experiments in the *Naval Institute Proceedings* and sent Hunsaker a signed complimentary offprint. In contrast to the previous year's cruise, the squadron stayed
close to home, shuttling back and forth in the Chesapeake Bay and spending
several weeks on Long Island Sound and the Maine coast before returning to
Annapolis on 28 August.[21]

Hunsaker graduated in 1908 at the head of his class, receiving top grades in
navigation, naval construction, experimental engineering, electrical engineering,
and seamanship. The school yearbook, the *Lucky Bag,* misspelled his middle
name as *Clark* but otherwise hit the mark with praise for his strong sense of
humor and "good share of common sense." The entry added that although he
"loves an argument," he "generally proves he's right."[22]

Midshipmen at the time did not receive their commissions immediately upon
graduation, usually serving a year or so on active duty before promotion to lieutenant junior grade. Not long after leaving the academy, and before the custom-
ary year of sea duty, Hunsaker determined to go directly into the Construction
Corps, the branch of the service that best suited his interests in engineering and
technology. While he was home in Saginaw, he mentioned his idea to his father,
who wrote to Secretary of the Navy Truman Newberry (a Michigan native and

Midshipman Hunsaker, 1908. He graduated from Annapolis first in his class. (Courtesy JCHP/MIT)

a friend), both U.S. senators from Michigan, and Congressman George A. Loud (another friend), asking them to intervene on his son's behalf with Adm. Washington L. Capps, chief of the Bureau of Construction and Repair (BUC&R). He received the same answer from everyone: Selections to the corps were made only after at least one year of service at sea.[23]

Hunsaker then traveled to Norfolk, where on 15 August he received orders to report for duty as a junior division officer aboard the armored cruiser *California* in San Francisco. Commissioned only four years earlier, the ship displaced 15,000 tons fully loaded and mounted a battery of four eight-inch guns and fourteen six-inch guns, in addition to an array of smaller-caliber weapons. The *California* cleared the Golden Gate on 24 August, bound for Pago Pago in American Samoa. When the ship crossed the equator, King Neptune and his minion, Davey Jones, initiated Hunsaker and the other neophytes into their realm. "There was a great show of outlandish costumes," Hunsaker wrote, "including the Barber with a great bucket of suds, a giant wooden razor and the final big canvas tank for immersion to wash it all off." Pago Pago, Hunsaker thought, was "lovely," with its wooded hills and friendly natives; one local family presented him with richly decorated wall hangings and a beautifully crafted ceremonial wooden bowl. And he could not help but notice one of the female members of the family, an eighteen-year-old who was "very pretty and well formed" and "wild to come aboard our ship to see how we lived." Heading back across the Pacific, the *California* put in at Hawaii, where Hunsaker and his shipmates swam off Waikiki Beach in Honolulu and danced with the daughters of the island's rich and famous at the Moana Hotel.[24]

At the end of October the *California* put in at Magdalena Bay, on the west coast of Mexico, and the next two weeks were spent in gunnery practice. Hunsaker was impressed by Magdalena Bay as a site to do a warship's business:

> It affords an excellent anchorage and is ideal for the purpose of target practice. . . . No shipping is in the way of great gun practice and the shore is absolutely devoid of valuable property. The place, in addition, affords no diversions nor visitors to interfere with drilling the men for target practice.[25]

In early December Hunsaker and the *California* were off to show the flag along the west coast of Central and South America. Hunsaker wrote a detailed narrative of this leg of the voyage, which appeared in serial form in the *Saginaw Courier-Herald*. Informative and folksy, the account included colorful descriptions of the people, social life, animals, and scenery. He complained that none of the attractive young women he met spoke English and lamented that his Spanish, although adequate enough "to buy things, engage rooms at a hotel or order a dinner," was too poor to allow him to converse with them.

The USS *California*. The armored cruiser took Midshipman Hunsaker down the west coast of South America and across the Pacific in 1908–9. It proved to be the only sea duty in his long navy career. (19N-11478, Courtesy National Archives)

In Chile he ventured into the high cordillera on a hunting expedition, learned to like garlic, and gained an appreciation of the native wines. (Hunsaker relished the occasional drink and later in life became something of a wine expert.) From Chile it was on to Callao and Lima, Peru, where he saw a bullfight and marveled at the mummified remains of the great conquistador Francisco Pizarro. The ship reached the Galapagos Islands in the middle of February. Unlike modern tourists to that strange and remote archipelago, Hunsaker was more interested in dispatching the local wildlife with a shotgun than in trying to preserve it on film, but like visitors today he found the huge tortoises fascinating just to watch. At night aboard ship he devised a rig to catch flying fish, which tasted "as good as fish ever get." By the third week of March the ship was back at Magdalena Bay to hone the crew's artillery skills.[26]

In charge of one of the ship's six-inch-gun batteries, Hunsaker applied his technical and analytical skills to the age-old challenge of aiming a ship's guns while compensating for the vessel's pitch and roll. Treating the problem as a three-dimensional mathematical exercise, he made a series of observations and computations, discovering that the gun muzzle traced a pattern of movement corresponding to the amplitude of a longitudinal wave and that exact calculations could be made to adjust the sight to the movement of the ship. He then fabricated an ingenious device that simulated the ship's movement and that provided more realistic training than was available with conventional methods. Praising him for his work on the ship's gunnery systems, the *California*'s commanding officer, Capt. Vincendon L. Cottman, reported that Hunsaker was a "very fine and able young fellow," who "gives every promise of making a valuable man for the service." Hunsaker, meanwhile, never wavered in his determination "to be a designer and an engineer" and renewed his quest to join the Construction Corps.[27]

Hunsaker was never entirely comfortable with the navy's belief that its line officers must be generalists, jacks-of-all-trade whose knowledge was a mile wide and an inch deep. In the new technologically oriented navy, bright young officers had open to them the prospect of more rapid advancement in engineering, with the proviso that once they entered the specialty they were not likely to command ships at sea. Hunsaker again asked his father to intervene on his behalf, and Walter went to Washington in January 1909 to pull strings. There he found a more sympathetic audience. After meeting with Newberry and having breakfast with Michigan Senator William A. Smith, Walter finally received assurances that his son would be able to join the corps without having to spend further time at sea. On 9 March 1909 the department issued orders assigning young Hunsaker to the Boston Navy Yard, with a concurrent appointment to the graduate program at MIT. He did not receive the orders until 13 April, just as the *California* was leaving Magdalena Bay, and it was not until his ship arrived in San Francisco a week later that he was detached and could make his way east. He arrived in Boston on 29 April. He and two other recent academy graduates assigned to the corps, Harry G. Knox (class of 1906) and Ralph T. Hanson (class of 1907), found rooms in a boardinghouse on Westland Avenue recently vacated by two officers who had just completed the MIT program.[28]

When Hunsaker joined another dozen Naval Academy graduates at MIT in the fall of 1909, the institute was jammed into a hodgepodge of old buildings in the Copley Square neighborhood of Boston's Back Bay. The institute had been chartered in 1861 but did not graduate its first class until 1868. After some lean years, the school expanded and, in the latter part of the nineteenth century, began to flourish, earning a national reputation for excellence in the physical sciences, mathematics, and engineering. It was designated a land-grant school under the 1862 Morrill Act, strengthening its ties with business and industry. It also had links to the military, for the Morrill Act mandated that the curriculum include courses in military science and that students undergo regular drill.[29]

Toward the end of the century, MIT had formed a symbiotic relationship with the navy, which was ever mindful of the need for engineers at the dawn of the new technological age. In 1898 Cecil H. Peabody, a professor of mechanical engineering, broached the idea of creating a three-year course in naval construction for recent Naval Academy graduates. The formal offer, however, was turned down by the navy on the grounds that the school did not have the facilities or the personnel to support the program. Peabody persisted and in 1900 submitted a revised plan, which the navy accepted. The school adopted the plan the following year. Those selected for the program spent their final two undergraduate years taking a course in naval architecture, followed by a year of graduate studies leading to a master of science degree. To head the program, MIT brought in William Hovgaard, an officer in the Danish navy and an internationally recognized ship designer.[30]

The curriculum for the naval architecture course included courses in ship design, mechanics, metallurgy, sanitary engineering, and electrical engineering. Hunsaker thrived in the academic environment at MIT, where he found the stress on original thought and creative reasoning more agreeable than the constrained intellectual atmosphere of the Naval Academy or the dull routine of life aboard ship. He also enjoyed the social life. In 1910 he joined the Delta Kappa Epsilon fraternity and for a time lived in the fraternity house on Newbury Street. Midway through his first year, Hunsaker came up for promotion. He passed the written examination on 14 March 1910 and on 1 April received his commission as a lieutenant junior grade, with duties as an assistant naval constructor.[31]

Soon after entering the MIT program in the fall of 1909, Hunsaker attended a tea at the Navy Yard commandant's house, where he met Alice Porter Avery, a young art student. A year his junior, Alice was the second of three daughters born to George W. and Elizabeth Porter Avery. Avery, a medical doctor with a private practice in Hartford, Connecticut, had died of pneumonia in 1893, driving his grieving widow into a deep depression, which led eventually to her hospitalization. For all intents and purposes orphans, the three Avery children wound up with their uncle, Robert Keep, who ran the prestigious Free Academy in Norwich, Connecticut. Alice attended Miss Porter's School, founded by her Aunt Sarah in Farmington, Connecticut, and run by the Keeps after Sarah's death. During their time with the Keeps, a friend of the family had financed a year in Europe for Alice and her older sister.

"After some maneuvering," Hunsaker learned that Alice now lived in Boston with two of her aunts, was studying painting at the Boston Museum of Fine Arts, and had a telephone: "A five-cent call seemed indicated," he remembered. Alice returned the affections of the "young Hunsaker," as she referred to him much later in life. Their courtship continued through the winter and spring of 1910, culminating in an engagement toward the end of the summer. The two married on 24 June 1911 at the old Porter homestead in Farmington, with Hunsaker's Annapolis roommate, Robert S. Young Jr., standing in as best man. Hunsaker took five weeks' leave, during which he and his new wife spent their honeymoon first at a cottage owned by the Keeps at Wauwinet on Nantucket Island, touring Algonquin Park in Ontario, and then visiting Saginaw.[32]

Back in Boston, the Hunsakers found an apartment at 1677 Beacon Street in Brookline and settled in with Pat, an English bull terrier Jerome had given to Alice in the summer of 1910. Although the apartment was small, the Hunsakers lived comfortably, well off enough to buy new furniture and to hire a maid to help with the housework. On 17 October 1912 Alice gave birth at home to their first child, Sarah Porter. A doctor and Alice's older sister, Mary, attended the birth. Sarah, nicknamed "Sally," unknowingly caused Pat's premature departure from the family. Instinctively protective of the infant, Pat took on all canine adversaries when the family went on walks with the baby. Fortunately, Hunsaker found Pat a good home with one of his friends at the Navy Yard.[33]

Hunsaker's marriage and honeymoon in the summer of 1911 interrupted work he had begun at the Boston Navy Yard on Taylorism, or scientific management. The brainchild of a Philadelphia engineer, Frederick Winslow Taylor, scientific management used time-motion studies to develop instruction sheets for basic tasks on the shop floor. Taylorism's most important consequence was to shift the responsibility for running factories from the shop foremen to a new cohort of professional managers. The system appealed to mechanical engineers in general and to Hunsaker in particular, who appreciated its rationality and promise of reform and efficiency. Moreover, models were not far away. In 1909 the Watertown Arsenal on the Charles River, outside Boston, instituted the Taylor system as a means of increasing the efficiency of the workforce and maximizing the output of the arsenal's machine shop.[34]

Rear Adm. Richard M. Watt, chief of the BUC&R, directed the Navy Yard to experiment with Taylorism in the manufacture of chains. Accordingly, Hunsaker visited shops in Quincy, Massachusetts, Schenectady, New York, Cleveland, and Philadelphia, where Taylor system experiments were under way. In Philadelphia the Link-Belt company had fully implemented the system and was a model for all converts to Taylorism. Hunsaker studied and reported on the arrangement of machine tools, the flow of materials, and the numbers and locations of shop-floor workers. He also got what he termed a "heavy dose" of scientific management at the Navy Yard, whose shop superintendent, Lt. John E. Otterson, advocated the system with "missionary zeal."[35]

Attractive as Taylorism was to Hunsaker and others in the navy's new technical elite, it did not mesh with the social makeup of the Watertown Arsenal and the Boston Navy Yard. Workers strenuously resisted the Taylorists' efforts to

Alice Porter Hunsaker, 1950. (Courtesy Alice Hunsaker Bird)

adapt the rational organization of objects and tools to the structure of people and their work, viewing the professional managers' time studies as disguised means of instituting a speedup. At the same time, foremen defended their autonomy on the shop floor and balked at the usurpation of their hegemony by the new "experts." At Watertown, opposition to the system manifested in a strike in 1911, a congressional inquiry, and the subsequent prohibition of Taylorism in government-run manufacturing facilities. Hunsaker liked how Taylorism removed the "interference" of the shop foremen and enhanced the power and control of the professional engineer in the workplace. Admitting he had been one of Taylorism's strongest adherents before moving away from management and concentrating on an advanced engineering degree, Hunsaker saw scientific management and the efficiency movement as integral to technological progress and social betterment.[36]

For his master's thesis at MIT, Hunsaker collaborated with Ralph Hanson in 1912 on a study of the twisting moment on a ship's rudder at high speeds. By comparing the results of their experiments with existing hydrodynamic data, the pair hoped to develop new formulas that could be used in steering-gear design. Hunsaker and Hanson received approval from their adviser, Cecil Peabody, and from the commandant of the Boston Navy Yard to fit instruments to the rudder of the destroyer *Sterett*. Less than two years old, the three-stacker had made more than thirty knots on its builders' trials, so there was little question that it would provide the performance that Hunsaker and Hanson were looking for. The pair designed and fabricated a recording dynamometer to measure the forces acting on the rudder and a helm-angle indicator to monitor rudder positions. In June they spent two days at sea as the ship went through an exhaustive series of maneuvers. The tests with the ship steaming ahead went as expected, but in two trials in which the ship ran astern at maximum speed so much water poured over the deck that the helm-angle indicator shorted out and the severe vibrations shook firebricks off the walls of the ship's boilers.[37]

Following acceptance of the thesis by the engineering faculty, Hunsaker presented his findings in a paper to the general meeting of the Society of Naval Architects and Marine Engineers in New York in November 1912. Published in the society's *Transactions* as "Rudder Trials, USS *Sterett*," the paper presented new information and demonstrated Hunsaker's solid grasp of empirical methodology and comprehensive understanding of hydrodynamics. Moreover, it had immediate practical application to naval architecture, especially as ships grew steadily larger and more powerful.[38] Most important, the study brought Hunsaker closer to a theoretical understanding of fluids in motion and pointed him toward serious thought about aerodynamics, which occupied him in one way or the other for the remainder of his long professional career.

2.   Aeronautics

Aeronautics was a burgeoning field in the early 1900s—comparable in many ways to electronics and computers at the end of the century. It is not surprising, then, that it attracted someone with Jerome Hunsaker's technical proclivities. While he was serving on the *California* in the fall of 1908, Hunsaker read about Orville Wright's flights at Fort Myer, outside Washington, D.C., which were marred by Orville's devastating crash and the death of army Lt. Thomas Selfridge

At the time, Hunsaker could see no practical use for the airplane and dismissed the Fort Myer demonstration as "just the Army trying to get some substitute for the horse" in its cavalry units. "What the fellow would do up in the airplane all that time, we didn't know and didn't care much. We had no concept that naval aviation would ever mean anything to us." Hunsaker's attitude changed in September 1910, when he attended the Boston-Harvard meet and saw flights by such pioneer aviators as Claude Grahame-White, Earle L. Ovington, and Henry H. ("Hap") Arnold. Many years later, he looked back at the Boston exhibition as the seminal episode that turned his attention to aviation and aerodynamics: "Considering the climate of the times, and that airplane flying was new, I wondered what kept them up, what made them steerable, what made them go out of control. So some study of aerodynamics would be a logical first step from the study of general dynamics."[1]

Hunsaker's new bride was instrumental in pointing him toward aeronautics. Some time before her marriage, Alice Avery became friends with Margaret Maclaurin, the wife of the new MIT president, Richard C. Maclaurin. Alice

Avery and Margaret Maclaurin were years apart in age, but the two women shared common interests and thought of each other almost as sisters. After Alice married, her relationship with Margaret Maclaurin became, if anything, even closer. The Hunsakers spent much of their time at the Maclaurins' house and even went on vacations with them. Hunsaker had great respect for Maclaurin, describing him as a "quiet man" who enjoyed the outdoors and had a strong sense of humor. He recalled that when Maclaurin was in the right mood he could turn on his Scottish brogue and mimic Harry Lauder, to the amusement of his friends and family. In short, the Hunsakers soon became part of both the Maclaurin and the MIT families.[2]

To fulfill his vision of a modern polytechnic university, Maclaurin decided to abandon the institute's cramped quarters in Boston and move to a new location, across the Charles in Cambridge. Maclaurin saw aerodynamics, a field in which Europeans had made rapid advances in recent years, as one of the priorities for the "New Technology." At MIT there had been scattered interest in the subject since 1896, when a master's degree student, Albert J. Wells, had studied changes in the pressure of air on flat, square surfaces using an open-circuit wind tunnel with a three-by-three-foot (0.91-by-0.91-meter) test section. In 1909 the engineering professor Henry A. Morse traveled to Europe to look at aeronautical courses in France, Germany, and England. Meanwhile, a group of students that included Frank W. Caldwell, who later introduced a practical variable-pitch propeller, formed the Tech Aero Club, which in the fall of 1909 concentrated on building and testing a monoplane glider and later a Curtiss-type pusher biplane.[3]

Student and faculty enthusiasm for aeronautics at MIT convinced Maclaurin that it was time to establish the subject as part of the curriculum. On a trip to England in the summer of 1910, he discussed the possibility of creating an aeronautical program with a friend, Richard T. Glazebrook, head of the prestigious National Physical Laboratory, which operated a four-by-four-foot (1.2-by-1.2-meter) wind tunnel at its facilities in Teddington, outside London. Maclaurin's ideas struck a resonant chord with Cecil Peabody, who thought that MIT's current course in naval architecture and marine engineering was a good starting point for a four-year program in aeronautical engineering. With Maclaurin's support and encouragement, Albert A. Merrill, an aviator and founding member of the Boston Aeronautical Society, presented a series of lectures at the institute in the spring of 1913, attracting upwards of ninety students. Following up on a recommendation from the Alumni Council, the Executive Committee of the MIT Corporation called for the construction of a wind tunnel and the creation of a graduate course in aeronautics. In December 1913 the committee set aside $3,500 for an aeronautical engineering laboratory.[4]

By this point, Hunsaker had been poring over the aeronautical literature in MIT's library, familiarizing himself with the work of Otto Lilienthal, Octave Chanute, Samuel P. Langley, and Frederick Lanchester. He also discovered a number of potentially important aeronautical works in French that had not been translated into English. In the June 1912 issue of the *United States Naval Institute Proceedings,* he published a translation of E. Lapointe's 1910 article, "Aviation in the Navy," which illustrated Lapointe's (and Hunsaker's) understanding of the mathematical foundations of the study of aerodynamics and the conviction that the future design and construction of aircraft had to be based on scientific principles.

Lapointe reviewed the formulas used for studying the center of pressure on surfaces moving through fluids and how they could assist in understanding the stability of airplanes. He showed, for example, how a combination of scientific studies and empirical analysis led aircraft designers to employ means of dampening longitudinal oscillations caused by shifts in the centers of pressure and gravity. He also examined the problem of lateral stability, compared the performance of biplanes and monoplanes, and demonstrated how mathematical analysis could help in the design of propellers. As for the potential of aviation at sea, Lapointe saw the airplane and airship playing vital roles in scouting and spotting and in antisubmarine warfare, and he recommended the development of specialized types to fulfill those missions. More than anything else at this early stage of Hunsaker's work, the Lapointe translation focused his thought on methods of mathematically ascertaining aircraft stability early in the design process.[5]

The Lapointe article led Hunsaker to a much more ambitious undertaking—the translation of Gustave Eiffel's classic book on aerodynamics, *The Resistance of the Air and Aviation.* Eiffel was a gifted French engineer who early on had recognized the advantages of wrought iron and had specialized in the construction of railroad bridges combining exceptional strength and aesthetic qualities. The internal framework of New York's 1884 Statue of Liberty was an Eiffel design, but he earned everlasting fame for the tower bearing his name erected on the Champ de Mars for the 1889 Paris Exposition. The massive yet delicate-appearing wrought iron structure, the first in the world to exceed 300 meters (984 feet) in height, was a spectacular expression of the engineer's vision and a symbol of nineteenth-century technological progress. In a tribute to Eiffel in 1919, Hunsaker referred to the tower as a "marvel of rational design."[6]

Beginning in 1902 with drop tests from his tower, Eiffel carried out over the next eight years a systematic program of aerodynamic experiments using wind tunnels at his laboratories on the Champ de Mars and at Auteuil outside Paris. Not the least of Eiffel's contributions was the popularization of the term *wind tunnel* and the introduction of the polar diagram used by engineers today to chart lift and drag coefficients. He also firmly established the principle of similitude. By comparing the data from the objects dropped from the tower with those from his wind tunnel experiments, Eiffel provided the first conclusive proof that the aerodynamic forces on a stationary object in a flow of air were the same as those on an object moving through still air. In addition, Eiffel's studies of airfoils demonstrated that lift can be calculated from the total of all the pressure distributions on a surface. Finally, Eiffel was the first to test complete airplane models in his wind tunnels, providing information for determining how much power was needed to achieve a given level of airplane performance. His work culminated in 1910 with publication of *The Resistance of the Air and Aviation*. Some consider the book "a masterpiece of clear engineering exposition, careful experimentation, and unique creative thoughts about experimental aerodynamics" that is "unappreciated" by today's aeronautical engineers.[7]

To Hunsaker, Eiffel's book was a revelation. Here in one volume was the theoretical foundation for the systematic, rational study of the forces at work on an airplane in flight, and it needed a wider audience. Not usually given to impulse, he wrote to Eiffel early in 1912 to see if he had made any arrangements for having his book translated into English and published in the United States. With Eiffel's permission to make the translation and negotiate its publication, Hunsaker canvassed reputable figures in aeronautics to see what they thought of the work and of its potential value to the field. Earle Ovington, who had flown a Blériot at the Boston-Harvard meet, responded enthusiastically. Eiffel's work, Ovington believed, "should be available to every one who is at all interested in the study of aeronautics," and its translation "would form the most valuable addition" to the literature.[8]

Finding a publisher proved difficult. Captain Washington I. Chambers, who had responsibility for aviation in the Office of the Secretary of the Navy, wrote that there would be some demand for the translation among aeronautics clubs, colleges and universities, and public libraries but that the total market was likely to be too limited to attract many publishers. Chambers's warnings about the potential marketability of the translation proved correct—at least initially. Hunsaker's inquiry to John Wiley and Sons in New York, who specialized in scientific and technical publications, brought a polite but firm refusal. Finally, an editor at Houghton Mifflin in Boston agreed to contact Constable and Company in London to see if it was interested. One of Constable's editors informed Houghton Mifflin that the publisher had checked with experts who were familiar with Eiffel's work and that they would publish the book provided that Eiffel and Hunsaker bought the first two hundred copies. Houghton Mifflin agreed to handle all sales in the United States.[9]

For more than two months Hunsaker spent nearly all his spare time working on the translation, making corrections, and arranging for the plates and tables to be sent to Constable directly from France. He was anxious to complete the work before Eiffel published any new information based on his continuing experiments. Well ahead of schedule, Hunsaker shipped off the manuscript to Constable on 23 October, including a brief translator's note and a list of Eiffel's works and urging them to expedite publication. In the meantime, Eiffel sent Hunsaker copies of a lecture he had given summarizing his most recent work at Auteuil. Hunsaker agreed to translate it and to forward it to Constable for inclusion as an appendix, even though he knew it would cause a delay. Hunsaker completed the tedious process of checking the proofs and returned them to Constable before the end of February. Eiffel was pleased with the book when it came out in June 1913, richly bound and presented in quarto size. The selling price of ten dollars reflected the high quality of the book's production.[10]

Reviews of the book were positive. The British journal *Aeronautics,* stating that Eiffel was "probably the greatest authority existent" in aerodynamics, affirmed that the translation will be of "immense service" to those doing research in the field. In the *Aeronautical Journal,* another British publication, Frederick Handley Page, a young aeronautical engineer, considered the translation "good," while suggesting a few changes in wording to make the meaning more precise. Handley Page particularly liked the book's supplement, which provided the first information in English on Eiffel's wind tunnel at Auteuil. In the United States, Chambers applauded the book as "the most valuable information on Aviation yet published," certain to be much sought after by designers, manufacturers, and engineering students.[11]

In the meantime, Maclaurin's plans for an aeronautical engineering program at MIT coincided with the efforts of Chambers, who wanted aeronautics placed on a firm scientific footing and who in 1911 had initiated a campaign for a national aeronautical laboratory paid for and operated by the Smithsonian Institution. David W. Taylor, then a captain in charge of the model basin at the Washington Navy Yard, opposed the idea, fearing that it would overlap with the work of the Bureau of Construction and Repair. He wanted his bureau and not the Smithsonian to be the leader, believing that aerodynamic research should be on a practical, engineering level rather than the more theoretical approach it would receive if the navy were not directly involved. Maclaurin was in a bind: He did not want to do anything that would offend the navy or Taylor; nor did he want to give the appearance of opposition to advances in either the theoretical or the practical understanding of aeronautics. In August 1912 he declared his support for the laboratory, without committing himself to either the navy or the Smithsonian and broadly hinting that MIT would be the best site for the facility.[12]

In 1913 the Smithsonian and MIT went their separate ways on the issue of the aerodynamics laboratory. Generally following the outlines of Chambers's proposals, Charles D. Walcott, secretary of the Smithsonian, reactivated the old Langley Aeronautical Laboratory, with an Advisory Committee making decisions about research projects and the allocation of money. At about the same time, Maclaurin determined that MIT should proceed with its own plans, reasoning that if and when a decision were made on a truly national laboratory, the institute's position would be strengthened if it had a program and facility already in place.[13]

Jerome Hunsaker was the key ingredient in the recipe. Confident that the young engineering officer could do the job and that his appointment would cement the alliance with the navy, Maclaurin met with Hunsaker on 22 April to see if he was interested in being detailed to MIT to develop an aeronautics curriculum and related research programs. Of course, he was. Hunsaker immediately wrote to Rear Adm. Richard Watt, the chief of the Bureau of Construction and Repair, outlining his plans for a systematic study of aeronautics at MIT and reminding him that this was an "exceptional opportunity" for the navy. "I should find the duty congenial," he added, "and would bring my full energy and enthusiasm into the problem."[14]

In June Secretary of the Navy Josephus Daniels approved Hunsaker's appointment to MIT for three years, starting immediately with a trip to Europe to survey aeronautical laboratories, followed by the initiation of an aeronautical engineering curriculum and the establishment of an aerodynamics research program. Now Chambers was in an awkward position. Sympathetic to Walcott's initiative, he did not want MIT seizing the upper hand in the quest for a national lab, nor could he oppose such close navy involvement in the MIT program. The best he could arrange was to secure permission for Albert F. Zahm, a pioneering aeronautical engineer at Catholic University and the recorder for the Langley lab's Advisory Committee, to accompany Hunsaker on his European tour.[15]

Hunsaker's orders assigning him to temporary duty in Europe came through on 10 July 1913, and he and Zahm were off from New York nineteen days later aboard the North German Lloyd liner *Kronprinz Wilhelm*. In Paris the eminent Eiffel greeted Hunsaker like a long-lost cousin, offering to let him work in the Auteuil laboratory as long as he wanted to. Eiffel left a lasting impression. Decades later, Hunsaker described him as a "stocky, elderly Frenchman with a grey pointed beard, very closely trimmed, very neat." His patent leather shoes had a mirrorlike gloss. He ran his laboratory like a typical French business manager, "who came to his office and sent for people to tell him what he wanted to know, and he told them what he wanted them to hear; and he put on his hat and went home to lunch and didn't come back until about five o'clock." Hunsaker adjusted to Eiffel's odd habits and spent the next two weeks with him and his associates observing and participating in wind tunnel experiments with models of Farman and Blériot airplanes.[16]

Hunsaker and Zahm also visited the aerodynamics laboratory at St. Cyr on the outskirts of Paris, which had been founded the year before by aviation benefactor Henri Deutsch de la Meurthe. Along with a small wind tunnel and other facilities, St. Cyr boasted a 1,300-meter (4,260-foot) straight track. Electrically powered cars equipped with instruments ran down the track, providing data on velocity, lift, drag, and the movement of the center of pressure on full-scale wings and determining the power and efficiency of propellers. During tests of a Blériot monoplane on the outdoor track, Hunsaker and Zahm observed that the experimenters had a hard time compensating for the effects of the wind and were uncertain what advantages such experiments had over actual flight tests of an instrumented airplane.[17]

The next stop for Hunsaker and Zahm was the University of Göttingen, where Zahm had connections with Ludwig Prandtl. A quiet, humble man, almost legendary in the intimate world of aerodynamics, Prandtl steered his laboratory's work along the lines of his theories of boundary-layer flow and skin friction, creating a firm mathematical foundation for an understanding of induced drag on airplane wings. A modest affair, the Göttingen laboratory housed an innovative closed-circuit wind tunnel. In contrast to tunnels that exhausted directly to the atmosphere, the closed-circuit tunnel recirculated the air, lowering power requirements, providing a smoother flow, and permitting temperature and humidity control. Hunsaker's impression of the Göttingen laboratory was that, although it was run by "men of extraordinary ability," the physical plant fell short of what he had seen in France.[18]

From Göttingen, Hunsaker and Zahm visited the DVL (Deutsche Versuchsanstalt fur Flugwesen Luftfahrt), a new and exceptionally well-funded and well-equipped government laboratory at the Adlershof, outside Berlin. Over hearty German food, beer, and Rhine wines, the aerodynamic duo discussed with the Germans their structural testing and power plant development work. Like Paul on the road to Damascus, Hunsaker had a revelatory experience at the DVL. To him, the big facility was "an eye-opener, because they were doing full-scale experimentation and testing—with a flying field, test pilots, and airplanes that were better instrumented than at any other place in the world." Lighter-than-air developments were also high on Hunsaker's and Zahm's priorities.

While they were in Berlin they flew in the *Sachsen,* a Zeppelin airship making publicity flights over the city, and they saw the landing of the naval airship L.2 following a long flight from the Zeppelin works at Friedrichshafen. Much impressed, Hunsaker probed Zeppelin company officials and engineers about details of the airship's design and construction, only to find the Germans unwilling to reveal any information that might be of immediate use to the Americans and that the Friedrichshafen factory was off limits to foreigners.[19]

Hunsaker and Zahm had better luck getting information about other aspects of German aviation during their visit. In Berlin they found August von Parseval, a respected designer of kite balloons and nonrigid airships, to be especially forthcoming, even inviting them to dinner at his apartment in Charlottenburg. Later, one of von Parseval's assistants escorted the pair to Bitterfeld for a detailed inspection of Parseval's airships. Another side trip was to the field at Johannisthal, where the German airship line DELAG had a large hangar and the German army operated the Flug und Sportzplatz. In Hamburg, Hunsaker and Zahm spent what Hunsaker called "a pleasant afternoon" with Friedrich Ahlborn, who ran a small hydrodynamic research facility.[20] As they had in France, Hunsaker and Zahm found the Germans—aside from those involved with Zeppelin-type rigid airships—to be open about their aerodynamic work and willing to share their knowledge with the Americans.

After Germany it was on to England and the National Physical Laboratory

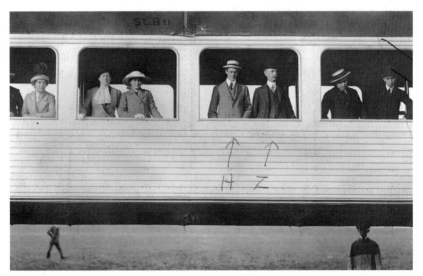

Hunsaker and Albert F. Zahm aboard the airship *Sachsen*, Berlin, 1913. (Courtesy JCHP/MIT)

in Teddington, where Hunsaker used Maclaurin's friendship with Glazebrook to good advantage. Glazebrook invited Hunsaker to work with the laboratory's Aerodynamics Group, headed by Leonard Bairstow. The principal concentration of the research was aircraft stability, which Hunsaker was familiar with from the Lapointe article, Eiffel's early investigations, and the work of Edwin B. Wilson at MIT. As a starting point, Bairstow and others relied on the work of G. L. Bryan of Cambridge University, which had shown that differential equations could be used to describe the six principal forces influencing an airplane in flight: lift, drag, side-wind forces, pitch, roll, and yaw. In wind tunnel experiments Bairstow ascertained the previously unknown coefficients in Bryan's equations. To Hunsaker the advantages were obvious: Aircraft engineers could now apply fundamental mathematical principles to design airplanes with greater or lesser degrees of stability and tailor the airplane to suit the specified purposes or mission, rather than finding out those characteristics after the airplane had been built and flown.[21]

While they were in England, Hunsaker and Zahm visited the Royal Aircraft Factory at Farnborough. An extensive facility with a large engineering staff, the factory did development work and testing in cooperation with the National Physical Laboratory and designed and built prototype aircraft that were put out to private manufacturing firms for series production. Leaving Zahm behind, Hunsaker went on to the Northampton Polytechnic Institute in London, where he gathered information on courses in mathematics, experimental aerodynamics, and airplane design.[22]

For Captain Chambers Hunsaker prepared a detailed, thirty-page report on airships, the only aircraft then available with the endurance, range, and payload capacity to fulfill the navy's demands for scouting and other fleet duties. From his report Hunsaker spun off a long article on airships for the *Journal of the Franklin Institute* in 1914. In this comprehensive and highly technical survey, he concluded that the Europeans enjoyed success in the field largely through the adoption of scientific methodology in airship design and construction. He weighed the advantages and disadvantages of rigid, nonrigid, and semirigid airships, examined their safety records (especially concentrating on the danger of fire presented by the use of hydrogen as a lifting gas), and speculated about their commercial and military applications. As to the future, Hunsaker foresaw continued development, particularly in the use of diesel power plants, which used less volatile fuel than gasoline engines, upgraded gas-tight fabrics, improved navigation instruments and techniques, and better weather forecasting.[23]

It is impossible to overstate the significance of Hunsaker and Zahm's European trip. For the first time, Americans had access to the latest aerodynamic research, firmly grounded in theory and mathematical principles. To be sure, the Wright brothers' invention of the airplane a decade before had been made possible by their aerodynamic experiments. Brilliantly conceived as it was, the Wrights' wind tunnel research had been empirical, intended to provide the data they needed to optimize their wing and propeller designs and not to add to basic theory. Before 1913 there was little attention paid to the subject of theoretical aerodynamics in the United States, the work of Taylor and Zahm being the sole exception. Hunsaker understood how fundamental this shift in emphasis from the practitioner to the theoretician was:

> In aeronautics, we in America are still in a transition stage. The preliminary research has been made and the fundamental invention produced. To perfect the first crude machines, more scientific research must be undertaken. We are at the point where the inventor can lead us but little farther, and it is to the physicist and the engineer that we must look for the perfection of air craft and the development of a new industry growing out of their manufacture and operation.[24]

The Hunsaker and Zahm trip was part of an intellectual transfer of major proportions, equivalent in some ways to that of the late eighteenth century, when Samuel Slater brought his knowledge of power textile machines to New England, or of the early nineteenth century, when French concepts of interchangeable parts stimulated a revolution in American machine-tool practice. Besides its significance for introducing European aerodynamic knowledge and methodology to the United States, the transfer established through Hunsaker a proprietary navy-MIT research and development partnership that eventually encompassed the National Advisory Committee for Aeronautics. Left out, the army eventually looked to Theodore von Kármán, one of Prandtl's former students at Göttingen who emigrated to Caltech in 1930 with his own agenda for the theoretical investigation of flight, as a means of creating an alternative research structure in competition with the navy-MIT-NACA axis.[25]

Hunsaker returned home aboard the White Star liner *Arabic,* arriving in Boston on 13 November 1913. Back at MIT, filled with enthusiasm for the new technology, he began fleshing out his ideas about a course in aeronautical engineering. Before the Alumni Council on 18 December, Hunsaker described how public interest in aviation had faded in recent years and observed that people were asking hard questions about what the new technology could be used for. Practical commercial applications for the airplane and the airship were not in the immediate offing. On the other hand, European military forces were directing their research and production programs toward the development of aircraft types for scouting and reconnaissance operations. Sooner or later, the United States needed to take a close look at aeronautics, which meant to Hunsaker a rational, scientific approach. "It is the function of our foremost engineering school to supply the technically trained men when they are required" through an aeronautical engineering course.[26]

In setting up a curriculum in aerodynamics, Hunsaker enjoyed the help of Cecil Peabody and Edwin B. Wilson. One of the most capable people on the faculty, Wilson taught in the Mathematics Department and had previously helped Hunsaker sharpen his thinking about the mathematical approach to the problem of heavier-than-air flight. The aeronautical engineering course was to be administered through the Naval Architecture Department, with Wilson taking primary responsibility for mathematics and theory and for an offering on dynamics and fluid mechanics. The bulk of the subjects in the course were Hunsaker's, among them the theory of airplane and airship design and practical drawing in those two subjects. The faculty and the MIT Corporation approved the degree program to begin in the 1914–15 academic year.[27]

Another outgrowth of Hunsaker's European trip was an article, "The Aeronautical Engineer," which appeared in the journal *Science Conspectus*. Here Hunsaker stressed how the British had rationalized their research in aerodynamics, with a top-level Advisory Committee for Aeronautics composed of some of the country's leading scientists and engineers coordinating the work of the National Physical Laboratory, the Royal Aircraft Factory, and other organizations. Other nations had made similar structural changes, with the result that the Europeans were in a race for "supremacy in the air." It was important, Hunsaker concluded, to understand that the competition was at all levels, not only aircraft production but also basic research and engineering.[28]

Integral to the MIT aeronautical engineering course was the laboratory Hunsaker helped to set up based on his observations during his European trip. Located in a wooden building on Vassar Street in Cambridge, the laboratory opened on 15 December 1914, earning the distinction of being the first facility on MIT's new site. Dominating the laboratory was an open-circuit wind tunnel, which closely followed the layout and performance of the National Physical Lab's first tunnel. The four-by-four-foot (1.2-by-1.2-meter) MIT tunnel employed a ten-horsepower direct-current electric motor linked to a four-blade propeller with a chain drive, generating speeds up to forty miles (sixty-four kilometers) per hour.

In addition to providing plans for the MIT tunnel, Glazebrook at the National Physical Lab arranged for the Scientific Instrument Company in Cambridge, England, to provide the new American tunnel's balance, a precision instrument used to hold models and other apparatus in the airstream and to transmit and

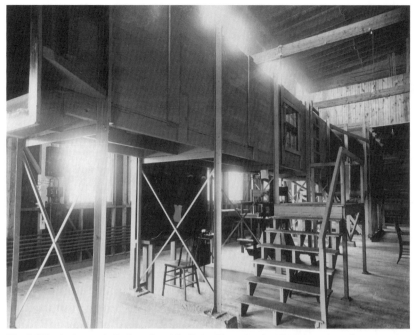

The MIT wind tunnel, designed and built by Hunsaker in 1914, based on the open-circuit design of a tunnel at the National Physical Laboratory in England. Hunsaker used
the MIT tunnel for his aircraft stability studies in 1915–16. (AA 320, Courtesy MIT
Museum)

record data. The MIT aeronautical laboratory was cold and drafty; one of Hunsaker's first chores was to stoke the fire in a pot-bellied stove in the often vain
effort to get the temperature in the building above 50 degrees F (10 degrees C),
essential for lowering the viscosity of the oil used to dampen the movement of
the balance. Later, when the institute's power plant opened next to the lab, the
crane used to move coal sometimes struck the side of the building, moving the
tunnel and jamming the balance, which rested on a concrete slab below the test
chamber.[29]

One of the first of its kind in the United States, the MIT program reflected
Hunsaker's thinking in effecting an equilibrium between science and engineering. Typically, students took classes in theoretical hydrodynamics from Wilson
and, from Hunsaker, classes in applied hydrodynamics, airplane design, airship
theory, aerial propellers, and laboratory work with the wind tunnel. If the success of a program is measured by the quality of its students, MIT did well. Among

those who came into the program in its inaugural year were H. K. Chow, who was the first to receive a degree in 1915, and Virginius E. Clark, an army officer who went on to a distinguished aeronautical engineering career. Notable, too, was Alexander Klemin, who became an instructor in the program.[30]

Even more well known was Hunsaker's assistant, Donald W. Douglas. Douglas had attended the Naval Academy for three years before resigning in 1912 to take the mechanical engineering course at MIT. He completed his degree in two years, only to encounter a depressed job market, and took the assistantship in order to make ends meet while he looked for more substantial employment. Hunsaker found him to be an intelligent and willing worker and was not surprised when he left MIT in 1915 to work for the Connecticut Aircraft Company in New Haven, where he helped design the navy's first airship, the unsuccessful A-1, and then to Cleveland to join the Glenn L. Martin Company as chief designer.[31]

Starting from the National Physical Lab's preliminary studies of dynamic stability, Hunsaker proposed a research agenda using the new MIT facilities. Bairstow had used an aircraft model to apply Bryan's stability equations, concentrating on longitudinal and lateral motions and discovering that they can combine to form complex oscillatory and nonoscillatory movements in flight. Hunsaker planned to use an inherently stable aircraft model in a series of wind tunnel experiments to examine these oscillations and to see if Bryan's and Bairstow's equations could be used to predict those motions. For the project he proposed studying six different aircraft wings, twelve biplane configurations, various fuselage shapes, and two complete aircraft models in what he estimated would occupy 750 hours of wind tunnel time. He began his work in the winter of 1915 and completed it in the spring of 1916, receiving his doctorate in June.[32]

In addition to his degree, the first in aeronautical engineering awarded by MIT, the principal fruit of Hunsaker's research was his *Dynamical Stability of Aeroplanes,* published in 1916 as part of the Smithsonian Miscellaneous Collections. Hunsaker employed $1/26$-scale models of two airplanes, one designed by Virginius Clark and the other a Curtiss JN-2 trainer. Clark's airplane had low wing loading (3.55 pounds per square foot [17.33 kilograms per square meter]) and a significant amount of longitudinal and lateral stability, while the Curtiss example had higher wing loading (5.2 pounds per square foot [25.38 kilograms per square meter]) and less stable characteristics. By applying Bryan's coefficients, Hunsaker developed his own data and performance curves for airplanes in equilibrium and in motion around three axes. The study's findings generally confirmed pilots' intuitive dislike of airplanes that were inherently stable. In horizontal flight, airplanes showed two types of longitudinal motion: the first was a slow undulating motion, the second a more rapid pitching movement. The undulating motion could be damped out at higher speeds using aerodynamic control surfaces, but at lower speeds it led to serious instability problems. Airplanes with augmented static stability, or "stiffness," demonstrated violent movements at slow speeds in gusty conditions, sometimes threatening total loss of control.

Lateral stability presented altogether different and more complex motions, typically combined as roll, yaw, and sideslip. The Clark airplane became seriously unstable at low speeds, occasioned by a rapid "spiral dive," or spin, under certain conditions, whereas the Curtiss model showed some of the same undesirable characteristics at higher speeds. Changes to the height and surface area of the vertical stabilizer and rudder, as well as to the degree of dihedral in the wing, greatly affected both airplanes' stability in normal turning and banking maneuvers.

Hunsaker concluded that the methods and data from his research were universally applicable and that, with "the systematic variation of one feature of the design at a time," there was no reason that a competent engineer could not make any airplane stable and safe in flight. "Radical changes" in design were not usually needed for most airplanes to achieve the desired degree of stability and control. Under combat conditions, an inherently stable airplane might have certain advantages over one that was not stable, because a pilot might be able to keep it in the air even though its control surfaces had been damaged. Concomitantly, as the data proved, an inherently stable airplane might become totally uncontrollable in gusty winds. Furthermore, wind tunnel tests were vital to the design process. It was difficult, if not impossible, to ascertain the effects of minor design changes by flight testing. Spin tests were dangerous, because the results could not be verified without repeating the experiment at additional risk to the airplane and its pilot.[33]

Hunsaker's stability research led to a technical report published by the NACA in its first annual report in 1915. Coming into existence as a rider to the naval appropriations bill of March 1915, the NACA was a compromise intended to defuse the lingering feud between the Smithsonian and the Washington Navy Yard over a national aeronautical laboratory. The NACA was a committee in every sense, its members chosen for their expertise in aeronautics and charged with oversight of existing government, university, and military aeronautical activities; it did not have its own laboratory or other research facilities. Hunsaker and Edwin Wilson received an $800 grant from the committee to prepare the report, the first in a prestigious series published over the next forty years. The report dealt with the effects of gusts on the longitudinal stability of the JN-2 airplane, highlighting the results of the wind tunnel experiments on the $1/26$-scale model. Widely distributed, the NACA's technical report provided valuable mathematical data for engineers concerned about predicting the stability of their designs before they entered flight testing.[34]

Before Hunsaker completed his research and published the results, Comdr. George C. Westervelt, a Construction Corps officer serving at the Puget Sound Navy Yard in Bremerton, Washington, approached him for help in the design of a twin-float biplane. In January 1916 Westervelt agreed to collaborate with William E. Boeing in Seattle on the machine, known as the B&W and the first in a long line of Boeing airplanes. One of Boeing's associates sent a $1/26$-scale model of the B&W to MIT, where wind tunnel tests by Hunsaker's team revealed that it had longitudinal stability problems. Minor modifications to the horizontal stabilizer provided an easy "fix." The work validated Hunsaker's mathematical models in predicting dynamic stability early in the aircraft design process.[35]

Hunsaker's stability study was an excellent early example of engineering science, especially when combined with his translation of Gustave Eiffel's book. The process had taken him from a general theory, expressed in abstract mathematical principles, through a series of scientific experiments, to a new theory that was universally applicable to aircraft design. Each airplane was essentially a package of technical compromises, its design filled with trade-offs of speed, stability, and handling, all of which had to be weighed against the anticipated mission of the airplane. Now firmly rooted in science and mathematics, rather than guesswork and trial and error, laboratory work in aerodynamics allowed the engineer to effect the best compromise at the least expense, before a design had been finalized and the airplane built.

Equally significant, Hunsaker's early aerodynamic work informs us about the nature of engineering knowledge. Walter Vincenti, an engineer who worked for the NACA in the 1950s, has used the history of aeronautical engineering to demonstrate that, although technology may apply science, it is not applied science. Rather, Vincenti maintains that it is better to think of engineering as design knowledge, whereby practitioners employ ways of planning, organizing, and structuring their work distinct from those of their scientific cousins. Moreover, to understand what engineers do it is also necessary to look at the difference between knowledge generated and knowledge used. Science tends to be knowledge generated (theoretical), and engineering is knowledge used (practical). But the dichotomy between the two is almost never that clear, with engineers often living comfortably in both worlds, simultaneously generating and using knowledge. Even though Hunsaker knew a great deal about the science underlying fluid flows, his aircraft stability studies at MIT in 1915–16 did not follow a straight line from scientific understanding to practical application. He began his work not with the desire to learn anything new about the fundamental nature of fluid flows but from the assumption that aircraft designers needed a common foundation of scientific and engineering knowledge about an important criterion of aircraft performance.[36]

As important as Hunsaker's pioneering work was in introducing aerodynamic theory to the United States, he might have made an even greater contribution if he had followed up on Prandtl's boundary-layer and skin friction research. Had he done so, Americans presumably would have had a deeper understanding of aerodynamic flows, induced drag, and the use of predictive mathematical models. Not until 1928 did major aspects of Prandtl's work appear in standard American engineering texts. A missed opportunity—yet in retrospect Hunsaker's decision not to replicate the Göttingen closed-circuit tunnel and follow up on Prandtl's work was understandable given Hunsaker's close ties with Eiffel and the immediate need for the solution of the vexing and dangerous problems of airplane stability.[37]

Hunsaker's years at MIT were busy and productive. He and Alice liked the Boston area; they were close to Alice's family, and they had many friends. During the summers of 1914 and 1915, they rented a cottage at Third Cliff in Scituate, just south of Boston. Among the many visitors who stayed with them were the Maclaurins and Hunsaker's parents, from Saginaw. In the spring of 1915 Alice became pregnant again, and the Hunsakers decided that it was time to move out of the apartment. They rented an old house at 25 Davis Avenue in Brookline, which despite its "antique" plumbing and heating had more than enough room. On 5 December 1915 Alice bore identical twin boys, James Peter and Jerome Jr., who like many twins were smaller than average at birth and required the care of a trained nurse for many months.[38]

But the expanding family was not to remain long in the Brookline house. On 16 June 1916 Hunsaker received word that he was to head the Aircraft Division in the Bureau of Construction and Repair. Hunsaker's orders sending him to Washington came through at the height of the nation's preparedness movement. Beginning in the fall of 1915, a wave of apprehension had swept over the nation that the war in Europe threatened the United States and that the country needed to bolster its defenses. Reluctantly at first, then with growing conviction and enthusiasm, the Democratic administration of Woodrow Wilson joined with a bipartisan majority in Congress in taking steps to build a "navy second to none" to enhance the nation's security. At the time of Hunsaker's posting to Washington, Congress was within weeks of approving a massive naval bill calling for the construction of ten battleships and six battle cruisers over the next three years.

Hunsaker had reservations about leaving Cambridge, which had been his home for more than six years and where he had made a major contribution to engineering science. He faced a daunting new task with vast responsibilities — reorganizing and expanding the Aircraft Division in anticipation of the needs imposed on naval aviation by the impending naval legislation. Nevertheless, the job allowed Hunsaker to apply his theoretical knowledge of aerodynamics to the real world of aircraft design and to bring his engineer's vision of order and efficiency to the navy's aircraft procurement system. It was the first step in his maturation as a heterogeneous engineer.

3.   The War in Washington

On 7 July 1916 Jerome Hunsaker reported to the Bureau of Construction and Repair in the State, War, and Navy Building, an ornate Victorian confection just west of the White House at the corner of Pennsylvania Avenue and Seventeenth Street. As head of the Aircraft Division, he relieved Lt. Comdr. Holden C. ("Dick") Richardson, who had been dispatched to Pensacola, Florida, to investigate a rash of training-plane accidents. Despite its important-sounding name, the Aircraft Division was a tiny appendage to the bureau, tucked away in a corner of the Battleship Design Office in Room 189, with no staff other than the secretarial pool.[1]

Hunsaker's boss, Rear Adm. David W. Taylor, had taken over as chief of the bureau in December 1914. Tall and distinguished looking, a man of great charm and intellect—possibly the most brilliant mind in the service at the time— Taylor had graduated from Annapolis at the head of the class of 1885 and had gone on to the Royal Naval College, where he had specialized in hydrodynamics. In reorganizing and strengthening the Construction Corps, Taylor sought the most skilled and enthusiastic young engineering officers he could find and instilled in them a strong sense of dedication and purpose. He quickly took Hunsaker under his wing, giving his twenty-nine- year-old protégé a sweeping mandate to expand the division's responsibilities to include all aircraft design, specifications, procurement, and inspection.[2] Along with other engineers at the time, Hunsaker viewed the impending conflict as an opportunity to demonstrate to the navy and official Washington how technical experts could organize a system capable of meeting the demands of the service for aircraft.

In putting together an organization and staff in the summer of 1916, Hunsaker discovered that only a handful of regular officers and civilians had the necessary skills and experience and that he could not strip aircraft manufacturers of the people they needed to accomplish their part of the job. Initially, a civilian, W. D. Clark, helped assemble a nucleus of draftsmen, rounding up people in the bureau who had worked on ships and instructing them in the nuances of aircraft draftsmanship. Hunsaker eventually found that he had to tap the ranks of reserve officers and draw on civilians for nearly all of his other assistants in Washington.[3]

It was a formidable task. In March 1915 Congress had appropriated the then-extraordinary sum of $1 million for the development of naval aviation, with the discouraging result that by the spring of 1916 the navy still had only a few satisfactory aircraft. An urgent requirement at Pensacola was for a front-engine training airplane to replace the old Curtiss "pushers," whose rear-mounted engines had a nasty habit of slamming forward in a crash and smashing hapless neophyte aviators. Hunsaker put together a procurement schedule for the new trainers only to realize that he could not follow the accepted competitive bidding procedures. On 11 August, after visiting some of the companies manufacturing navy aircraft, he prepared a pessimistic report for Taylor, noting that all of the contractors were having problems. Of two tractor-type airplanes Curtiss was working on, one was badly overweight and the other had been destroyed during trials. The Burgess Aeroplane Company, in Marblehead, Massachusetts, had done no better; the company was late in delivering the first of six trainers that had been ordered by the navy, and the airplane also exceeded weight specifications. Burgess promised that it would improve the situation, but Hunsaker thought the "prospect is poor" for the company to meet its contract obligations.[4]

Because "it was clear that the regular procedure of calling for competitive bids was too slow a process to meet the urgency of the situation," and knowing the trouble manufacturers were having with new designs, Hunsaker sought alternatives. He asked Taylor to secure permission from the chief of Naval Operations, Adm. William S. Benson, to acquire thirty trainers as proprietary articles, which meant that the navy did not have to go through the usual procurement process. To expedite matters, Hunsaker insisted that the airplane be a floatplane version of the successful Curtiss JN-4B army trainer. When wind tunnel experiments revealed that the JN-4 fitted with a single float could land and take off only at unacceptably high speeds, Curtiss had lengthened the wingspan, discovering in the process that the airplane exhibited serious stability and control shortcomings. Working with Curtiss, the Aircraft Division thoroughly redesigned the airplane as the N-9, with an advanced V-bottom float permitting safe low-speed takeoffs and landings, a longer tail, and "skid fins" on the top wings to enhance its control characteristics. On 1 September the navy signed a contract with Curtiss for the airplane, which the company completed within three weeks. By the end of the year, Curtiss and nine other manufacturers were building trainers for the navy, and the immediate crisis in Pensacola was over.[5]

More disconcerting to Hunsaker and the Aircraft Division was the acquisition of a small, high-powered shipboard floatplane known as the Speed Scout. Hunsaker and his staff completed the design for the airplane in the fall of 1916. He considered the requirements "to be difficult but not impossible" for the industry to meet, and the navy awarded contracts for four airplanes to Curtiss and Burgess. After an encouraging start, things began to unravel in 1917. By May, Curtiss had still not finished either of its airplanes, and the one example Burgess completed did not meet the division's performance requirements. Recognizing that the industry was seemingly unable to meet the navy's demands, Hunsaker revised the Speed Scout design in the spring of 1917, only to get a limited response from the industry. In desperation, the navy issued contracts to Curtiss and the Thomas-Morse Aircraft Corporation of Ithaca, New York, for sixteen Speed Scouts, even though they did not meet the Aircraft Division's original specifications.[6]

The episode convinced Hunsaker that the industry would have to change if it were to provide the kinds of aircraft the navy needed at a cost it could afford. "The Curtiss Aeroplane Company have developed a few types which are fairly acceptable and as a consequence have been given large production orders by the Army and Navy," he informed Taylor in May 1917. "No other company has developed equally acceptable types, and hence the other companies cannot be given orders of any magnitude." Hunsaker believed small manufacturers lacked the necessary engineering personnel, resources, and experience; at the other end of the spectrum, Curtiss, by far the largest manufacturer, had fumbled some of the navy's more specialized jobs and had been reluctant to bring its costs down. When Hunsaker told Clement M. Keys, a Curtiss vice president, that "unless he is prepared to show that his prices are reasonable, we ought to have an investigation of costs by Department accountants," Keys replied that his company had large overhead and development expenses and that the chances of cutting costs in the near future were slim.[7]

In large measure, the experiences with the N-9 and the Speed Scout in 1916 and early 1917 led Hunsaker to support the creation of a government-owned and -operated aircraft factory as a means of forcing the industry to cut costs and to meet the navy's requirements. In the summer of 1917 at Taylor's recommendation, Secretary of the Navy Josephus Daniels approved $1 million for the construction of the Naval Aircraft Factory at Philadelphia. Completed in record time, the factory became a major supplier of large patrol flying boats for the navy during the war.[8]

Hunsaker also knew that meeting the increased demand for airplanes stemming from the huge Naval Appropriations Act passed in August 1916 meant overhauling the navy's unwieldy inspection system, the responsibility for which was divided among the Office of Naval Aeronautics (under the Office of the Chief of Naval Operations), the Bureau of Construction and Repair, and the Bureau of Steam Engineering. In November 1916, Hunsaker wrote, "There is a very strong feeling among the flyers that this Bureau [of Construction and Repair] should not in any way originate, develop, or fix designs of aeroplanes for the service." Moreover, the aviators believed they were "the only persons competent to deal with the problem of the development of naval aircraft. They even go so far as to assert that though [H. C.] Richardson holds an expert license, he is not in their class because they have 'more hours in the air.'" Hunsaker concluded that the inspection system was deeply flawed and that if it were not soon changed an accident resulting in injuries or deaths would bring an investigation of the way the bureau went about its business.[9]

Hunsaker had a solution. In a long memorandum to Taylor on 13 December 1916, he highlighted the differences between the Bureau of Construction and Repair, which had primary responsibility for airframes, and the Bureau of Steam Engineering, which oversaw power plant design and procurement. Hunsaker suggested placing someone from the Bureau of Construction and Repair in the new position of superintending constructor of aircraft, with aviators posted to the factories to generate "good will and confidence" among the operating personnel and to be educated in the complexities of aircraft design and construction. Hunsaker also recommended that the Bureau of Construction and Repair assume sole responsibility from the Bureau of Steam Engineering for the inspection of engines and propellers. Taylor acted quickly on Hunsaker's recommendations to initiate reform of the navy's inspection procedures. Commander Westervelt, who Hunsaker knew from the aerodynamic experiments on the B&W floatplane in the MIT wind tunnel earlier in the year, became the navy's superintending constructor at the Curtiss plant in Buffalo.[10]

Hunsaker also sought to bring his and his colleagues' technical expertise to bear on the design and construction of lighter-than-air craft in the prewar years. In response to British reports emphasizing the potential of airships for maritime patrols, Capt. Mark L. Bristol, the head of naval aviation in the Office of the Chief of Naval Operations, recommended contracting with the Connecticut Aircraft Company in May 1915 for the construction of a nonrigid airship. When the company ran into trouble with the design, the navy turned to Hunsaker, then still at MIT, to provide wind tunnel data on stability and handling characteristics. Originally to have been delivered by October 1915, the airship, designated the DN-1, was far from finished by the summer of 1916. Disgusted with the situation, Hunsaker reported to Taylor in August that "the whole design is an amateur job," that the envelope had not yet been tested, and that the car, engine, and suspension system were not ready. Taylor believed that the service had made a mistake in relying almost entirely on an inexperienced contractor and ordered Hunsaker to begin studies of new airship designs.[11]

Although Hunsaker had studied and reported on European prewar developments and had taught courses on airship design while he was at MIT, he was not well versed in what had been done since the outbreak of war. Attacking the problem systematically, Hunsaker and his small but burgeoning staff assembled all the technical literature they could find. Nevertheless, there were still wide gaps in the Americans' knowledge. Hunsaker focused first on a small training airship, relying in part on information supplied by Lt. John H. Towers, who had just returned from duties in London as assistant naval air attaché with information about the British SS (Sea Scout) nonrigids. Then just as he thought he had the training airship design well in hand, Hunsaker received word from Benson on 13 December that the specifications were to be changed to provide for coastal patrol duties in addition to the training role. This meant that Hunsaker and his staff had to rethink the entire design in order to incorporate significantly enhanced range, endurance, and payload capacities.[12]

The airship project meshed perfectly with Hunsaker's ideas about the application of aerodynamic theory to practice. Because of the paucity of data on the design and performance of foreign airships, he and his team had to rely heavily on mathematical calculations, models, and wind tunnel testing. In keeping with his assumption that the initial airship design should be conservative, Hunsaker wanted the craft to be of the smallest size consistent with achieving the estimated performance. The availability of power plants also had a bearing on size. In 1916 only the Curtiss and Hall-Scott companies could supply engines that had the requisite power and reliability. Even though two engines would enhance the airship's performance, their fuel consumption was high, which meant larger tanks and a consequent increase in the overall size and weight of the craft. Caught in the aeronautical engineer's dreaded fuel-weight spiral, Hunsaker reluctantly decided to stay with a single engine. The final design was for an airship of about 77,000 cubic feet (2,180 cubic meters), with a total lift capacity of 5,275 pounds (2,393 kilograms) using hydrogen gas, an endurance of twelve hours, and a top speed of forty-five miles (seventy-two kilometers) per hour.[13]

Hunsaker completed the airship design and specifications by the end of the year, and Taylor forwarded them to the secretary of the navy for his approval in

January 1917. Originally, it had been intended to construct only one example of the airship, now known as the Type B to distinguish it from the still unfinished Connecticut Aircraft ship, which had been redesignated Type A. On 4 February, as the diplomatic and military situation continued to deteriorate and it seemed only a matter of time before the United States became involved in the European war, Daniels ordered the construction of sixteen of the new airships. To Hunsaker the directive was a "thunderbolt." There would be no time to evaluate a prototype; worse still, there was only one company—Connecticut Aircraft—with experience in building airships. Even if Taylor and Hunsaker had confidence in Connecticut Aircraft (which they did not), neither that company nor any others had the capacity for series production of that magnitude.[14]

Taylor determined that the best way to procure the airships was to sidestep competitive bidding and divide the order among Goodyear, B. F. Goodrich, U.S. Rubber, Curtiss Aeroplane, and Connecticut Aircraft, at the same time inviting them to send representatives to meet with Taylor, Hunsaker, and others in Washington on 12 February. To Hunsaker the conference, coming as it did amid heightened international tensions and rumors of war, "resembled a patriotic meeting rather than a gathering of prospective government contractors." The Curtiss people stated up front that they wanted to complete the entire order for the airship cars in their new engineering facility at Garden City, Long Island. They believed, too, that they could assemble all of the airships, provided they received timely deliveries of the rubberized fabric envelopes. Goodyear vice president Paul W. Litchfield, an MIT graduate and advocate of lighter-than-air craft, had no reservations about his company's capacity to provide as many airship envelopes as the navy needed.[15]

From experience with previous aircraft programs, Hunsaker and Taylor were not as sanguine about the manufacturers' prospects of meeting the navy's needs. Like any aeronautical engineer, Hunsaker was worried about weight and cautioned those present that unless they could bring the airships in at or below the weight specifications, the craft were "not worth building." Taylor warned that no one in the United States "had much experience in this line of work" and doubted that the prospective contractors could meet the navy's requirements. Only Goodyear, which had built balloons for the navy and had Ralph H. Upson, one of the country's most experienced lighter-than-air men, on its engineering staff, looked promising. Moreover, the company agreed to develop, at its own expense, a flying field and an airship hangar at Akron. At the end of the meeting there was general agreement that the manufacturers should create a committee to determine how best to share information, distribute the work, and pool resources.[16]

Goodyear received the bulk of the order when the contracts went out on 14

March—nine airships—with Goodrich getting two, and Curtiss three. Only after assurances from Connecticut Aircraft that it was trying to correct its engineering deficiencies did that company receive a contract for the remaining two. The detailed specifications called for a single-engine, two-man airship with a total volume of 77,000 cubic feet (2,180 cubic meters), within an envelope 160 feet (48.8 meters) long. Endurance was ten hours at a maximum speed of forty-five miles (seventy-two kilometers) per hour, increasing to sixteen hours at thirty-five miles (fifty-six kilometers) per hour. Delivery was to be within three months from the date of the contract. Curtiss eventually decided to turn its three airships over to Goodrich and worked out an agreement to manufacture all of the cars, engines, and aerodynamic surfaces for the fourteen airships to be built by Goodyear and Goodrich.[17]

Because its big hangar in Akron was not yet finished in May 1917, Goodyear shipped the first of its navy airships to Chicago for tests in a large Goodrich-owned hangar on the site of the old White City amusement park. Ralph Upson was so pleased with the airship's performance that on 30 May he decided to make an overnight flight from Chicago to Akron, providing dramatic proof of the soundness of the airship design. Goodyear delivered the B-1 to the navy in the middle of July and completed the remaining eight craft by March 1918. Goodrich finished its first airship in September 1917 and delivered the rest by June 1918. Even Connecticut Aircraft, whose work had done little to inspire the navy's confidence, delivered its two airships by April 1918. Hunsaker was pleased with the results of the B-1 program. The airships performed well in service, and the project had given him valuable experience with the contract procedures and policies he would need to rely on after the United States entered the war.[18]

In the months before American involvement in the war, Hunsaker was involved in the development of lightweight alloys for aircraft construction. Metal appealed to Hunsaker in part because it represented progress toward the industrialization of aircraft manufacturing, marking a break with the old craft tradition of making airplanes from organic materials. Moreover, metal construction was more "scientific" than wood construction for aircraft.[19] Yet because both wood and metal had inherent limitations—wood lacking homogeneity and dimensional stability and metal suffering from weight, fabrication, and corrosion problems—Hunsaker preferred an orderly, rational approach to the new material.

Much of the early information on lightweight alloys filtered in from Europe. When Hunsaker and Zahm toured Germany in 1913 they learned about the Germans' use of aluminum alloys in the construction of rigid airships, although the exact composition of those metals was a closely held secret. After the outbreak

Type B nonrigid patrol airship. Designed by Hunsaker and the Aircraft Division of the Bureau of Construction and Repair on the eve of America's entry into World War I, the airship proved useful in coastal antisubmarine patrols during the conflict. (A-3324-E, Courtesy NASM)

of war in the summer of 1914, information came from studies of German airship wreckage, but much of that intelligence was unreliable due to the haphazard way in which material samples were obtained and tested. Through Hunsaker, the Bureau of Standards acquired additional structural material scavenged from downed German airships. By the end of 1916, the bureau's tests showed conclusively that the principal alloy in German airship girders was duralumin, an aluminum alloy containing 3.0–5.0 percent copper, 0.4–1.0 percent manganese, and 0.3–0.6 percent magnesium. After careful heat treating, quenching, and age (or precipitation) hardening, the material had the tensile strength of mild steel.[20]

Not long after taking over the Aircraft Division, Hunsaker canvassed his many contacts in the engineering profession to see what they knew about aluminum alloys in airship construction. His friend Harry Knox, now a commander at the Norfolk Navy Yard, told him that he had done tests on a promising aluminum alloy supplied by an Ohio firm and that Vickers in Britain was using duralumin in airships. But the most encouraging word came from Alcoa, the nation's largest manufacturer of aluminum, whose engineers told Hunsaker that they would work with the navy to determine the composition of duralumin and install equipment to fabricate the material. Taylor agreed to let Alcoa proceed with the project on 17 August 1916, starting a long-term relationship between the navy and Alcoa that had major implications for cooperative military-industry research and development projects in general and for metal aircraft construction in particular.[21]

By the end of 1916 Hunsaker had provided Alcoa engineers with as much information as he had on light alloys and supplied small quantities of material for testing. In return, Alcoa agreed to supply the bureau with whatever samples and data it considered necessary. This arrangement led, by the spring of 1917, to 17S, an alloy that compared favorably with the German duralumin. Hunsaker, meanwhile, had found that his preliminary airship girder designs were deficient because he had not fully taken into consideration the unique characteristics of the new metal. Alcoa later took over girder design and fabrication along with its metallurgical responsibilities, but not until after the war did the company perfect rolling and stamping methods needed to shape duralumin for rigid airship construction.[22]

Now fully committed to the investigation of light alloys and their application to aircraft construction, the navy moved closer to the design and acquisition of a rigid airship. At its first meeting on 26 February 1917, the Joint Army and Navy Airship Board, chaired by Admiral Taylor and including Hunsaker, Towers, and Hunsaker's former student, the army captain Virginius Clark, decided that the Bureau of Construction and Repair should have primary responsibility for investigating rigid airship design and construction. In March the board determined to go ahead with a rigid airship design. Hunsaker and Clark estimated the cost would exceed $1.5 million and proposed a cooperative arrangement, with the army paying the salaries of a civilian engineering staff and the Bureau of Construction and Repair providing the office space and support personnel. The war and navy departments approved the plan on 2 April. To head the staff, Hunsaker recommended Starr Truscott, a naval architect and graduate of the University of Michigan who had practical experience with engineering projects at Newport News, Virginia, and the Panama Canal.[23]

Almost immediately after his arrival on 1 May, Truscott and his staff began investigating potentially promising airship designs. At this early stage of the process, Truscott gave careful thought to mounting the engines within the airship's hull, feasible only with the employment of nonflammable helium as a lifting gas. At a meeting on 26 July, the Joint Board approved a plan to have the

Bureau of Mines initiate limited helium production at a pilot plant paid for by the army and the navy, beginning what Hunsaker later referred to as "the boldest and most romantic enterprise of applied physics ever known to the world." It cost millions of dollars, but within a decade the United States had established an industrial infrastructure and gained a global monopoly on what at the time was one of the world's most valuable strategic resources.[24]

As Hunsaker and the Aircraft Division rationalized the navy's design and procurement process in the spring of 1917, what now seems inevitable occurred on 6 April, when the United States declared war on Germany. Provoked by unrestricted submarine warfare and the sinking of unarmed American merchant ships in clear violation of accepted international law and neutral rights, the Wilson administration and Congress saw no other choice than to take the nation into the European conflagration on the side of the Allies.

Unsure what course the conflict would take, Hunsaker and his staff prepared for a protracted struggle demanding exponential growth in the navy's demands for aircraft and aviation equipment. On 5 May the war and navy departments created the Joint Army and Navy Technical Board to develop plans and specifications for army and navy aircraft and to provide advice on procurement. Hunsaker, Towers, and Lt. Arthur K. Atkins from the Bureau of Steam Engineering represented the navy on the board, with the pioneer aviator Maj. Benjamin D. Foulois, Virginius Clark, and Capt. Edgar S. Gorrell sitting for the army. One of the problems with the board was that it lacked direction; Hunsaker noted a year later that the Technical Board's "function and authority have never been clearly defined."[25]

Without any policy statements or guidelines from either the navy or the war departments, the Technical Board took it upon itself to develop its own plans for the types and numbers of aircraft the nation's air services needed. The French ambassador in Washington passed on to the board information he had received in a cable from Premier Alexandre Ribot calling on the United States to supply, by the spring of 1918, 4,500 combat aircraft at the front, augmented by 2,000 replacement aircraft per month, and the requisite number of engines and spares. With a quantitative requirement, albeit from a foreign power and totally lacking in guidance as to what types of aircraft were needed, the board went ahead with plans for a huge expansion of American production. On 29 May the board agreed that American manufacturing capacity should expand to accommodate the manufacture of 10,000 training and 12,000 service aircraft.[26]

Another organization with responsibility for aviation was the Aircraft Production Board, formed on 16 May and headed by Howard E. Coffin, a vice president of the Hudson Motor Car Company. Taylor was a prominent member of the Production Board, which after October 1917 included two other navy officers—Capt. Noble E. Irwin, the new director of naval aviation in the Office of the Chief of Naval Operations, and Arthur Atkins (now a lieutenant commander). Hunsaker had nothing but praise for Taylor, who "by his complete grasp of the situation, his personal eminence, and his unfailing tact" clearly articulated the navy's aircraft requirements during the early phases of the war.[27]

The recommendations of the Technical Board and the Production Board set the navy's aircraft procurement wheels in motion. By September a consensus had emerged that large flying boats were essential to counter the immediate and future U-boat threat. In response, Hunsaker's Aircraft Division put together a proposal for the procurement of 1,700 aircraft, including 100 Curtiss H-16 flying boats powered by twin Liberty 300-horsepower V-1649 engines, 825 Curtiss HS-1 flying boats with single Liberty engines, and 640 Curtiss R-6 twin-float trainers.[28]

Not long after Hunsaker had presented his program, Adm. William S. Sims, the commander of American naval forces operating in European waters, sent an urgent request to the navy department calling for 630 more twin-engine flying boats, bringing the total to 730. Taylor was furious, telling Sims that his recommendation meant "the whole program is shot to pieces," but it was to no avail. Now Hunsaker had to plan for the production of nearly four times the original number of twin-engine flying boats, and to make matters worse, none of the existing plant facilities could cope with the additional requirement. The solution to the production dilemma was to assign 480 of the airplanes to the Naval Aircraft Factory, on the assumption that the factory would rely on outside contractors for major components and carry out final assembly only.[29]

All the aircraft in the Naval Aircraft Factory's new order were F-5-Ls, Liberty-engined versions of the British F-5, which looked like H-16s to the casual observer but which in reality were substantially different airplanes. It fell to Hunsaker's Aircraft Division, working in concert with draftsmen at the aircraft factory, to turn the British design into working drawings for the American F-5-L. On the face of it, it seemed like an easy task, but the British specifications and drawings were almost totally useless for the quantity production envisaged. As a consequence, Hunsaker's staff redrafted the British plans and specifications for the F-5 to make them suitable to American methods of manufacturing. The final product was identical to the British F-5 in appearance and external dimensions, although it had twin Liberty engines, a stronger hull reinforced with steel tubing, solid wing spars, a strengthened tail structure, and dual flight controls. The first F-5-L flew at Philadelphia on 15 July 1918.[30]

By the summer of 1918 Hunsaker was confident that the expanded flying boat program was headed in the right direction. Then he learned that Daniels wanted another 700 F-5-Ls plus an additional 600 HS-type and F-model single-engine flying boats to be delivered by 1 July 1919. Hunsaker and members of the Aircraft Division met in Irwin's office on 13 June to discuss Daniels's request. They decided to divide the F-5-L order among Curtiss Aeroplane, Curtiss Engineering in Garden City, the Naval Aircraft Factory, and a Curtiss subsidiary, Canadian Aeroplanes in Toronto. Three other firms, L-W-F Engineering in College Point, Long Island, Aeromarine in Keyport, New Jersey, and Boeing in Seattle were to receive orders for the single-engine airplanes.[31]

Not surprisingly given the magnitude of the program, the navy's 1918 flying boat effort encountered more than its share of obstacles. One was coordination of the activities of those firms building the same types of aircraft, each of which needed large quantities of parts and materials at roughly the same time. Materials shortages were a particularly vexing problem in the fall and early winter of 1917, with spruce topping the list. Fearing that shortages of West Coast spruce would endanger "the entire Navy aircraft program," Hunsaker identified new sources of wood in New England.[32]

Another problem Hunsaker faced was finding a contracting policy acceptable to both the navy and the manufacturers, who were sometimes reluctant to expand their plants without an understanding that the navy would insulate them from the financial risk. A way around this problem was to substitute the cost-plus-a-fixed-fee contract for the old flat-price contract, committing the navy to guaranteeing a profit of 15 percent based on a "bogey" price agreed upon by both parties. In addition, the navy contributed one quarter of any saving from the "bogey" price, thus encouraging the manufacturer to come in under that price.[33]

There was also the problem of getting the airplanes to the operating forces overseas. Navy flying boats had to be carefully crated and shipped to avoid damage. Typically, a special freight train, loaded with that week's production, left the Curtiss Buffalo plant every Saturday evening and headed for the docks at the Bush Terminal in Brooklyn, where the crated flying boats were put on board merchant ships bound for Europe. By early May 1918, thirteen months after the country entered the war, four steamers were on their way across the Atlantic with thirty-seven HS-1s and twenty-eight H-16s. Even someone as bright and resourceful as Hunsaker could not coordinate aircraft production to mesh with shipping schedules; quite by accident, production began to peak in the summer of 1918 at the same time more shipping space became available on the transatlantic routes. Luckily, too, flying boats, though bulky, did not weigh much and once crated could be loaded on the decks of many types of ships, including tankers, which usually could not accommodate much if any deck cargo.[34]

In conjunction with the expansion of the navy's flying-boat program came renewed emphasis on nonrigid airship production. In early 1918 Admiral Sims cabled the navy department to say that he needed more airships for antisubmarine duty in Europe. Within the Bureau of Construction and Repair the experience with the single-engine B-class nonrigids convinced everyone that an airship with enhanced range and payload capacity would be necessary for coastal patrols and to accompany convoys in the Atlantic. On 5 February 1918, representatives from the Bureau of Construction and Repair, the Bureau of Steam Engineering, and the Aviation Office of the chief of Naval Operations met to consider various design proposals, finally settling on a twin-engine 170,000-cubic-foot (4,814-cubic-meter) airship with a cruising speed in excess of fifty miles (eighty kilometers) per hour and an endurance of ten hours. They agreed that the British O-type airship, although it was a single-engine design, would be a good starting point, and they understood that the British would send over one of their engineers to help out. Thirty airships were to be manufactured, with deliveries of the first ten to start in early July and the rest of the production to extend into the spring of 1919.[35]

Determined to obtain the best performance possible, Hunsaker and his staff designed an entirely new envelope and car, using extensive wind tunnel experiments to minimize drag and maximize speed. Much attention was also given to improving the envelope fabric, with the goal of preventing its rapid deterioration, especially when exposed to intense sunlight. In their final form, specifications called for a 192-foot (58.5-meter) -long airship powered by two 125-horsepower Wright-Hispano engines, with a volume of 182,000 cubic feet (5,154 cubic meters), a total lift of 12,700 pounds (5,760 kilograms), and a useful capacity of 4,760 pounds (2,159 kilograms). Top speed was sixty miles (ninety-seven kilometers) per hour; endurance at forty-five miles (seventy-two kilometers) per hour was forty-seven hours and at fifty-five miles (eighty-eight and a half kilometers) per hour was twenty-eight hours.[36]

On Taylor's recommendation, contracts for the new ships, designated the C-class, went out in late May to Goodyear and Goodrich, both of which had proved themselves with the B-class airships. Goodyear received orders for six, Goodrich for four, and the Burgess firm was given responsibility for constructing all ten cars. Numerous changes to the design and specification, some initiated by Hunsaker's office and some suggested by the manufacturers, contributed to delays in meeting the original delivery deadlines. Not until the middle of August did Goodyear have the first airship envelope ready and Burgess get the first car to Akron. Goodrich, meanwhile, had yet to start work on any of its airships.[37]

Frustrated by the delays, Hunsaker went to Akron to see for himself what was holding up the C-class program. On 14 September, he reported that the

Goodyear airship was fully inflated and had been ready to fly for a week but that the first test flight had been postponed pending receipt of a few "trivial items" from the Bureau of Steam Engineering. He urged that "strong representations" be made with Steam Engineering to do whatever was necessary to get the parts to Akron. Despite Hunsaker's entreaties, more than three weeks passed before the C-1 took to the air.[38]

None of these programs was possible without a major expansion of the Aircraft Division's staff. Growing from only 14 people before the entry of the United States into the war, the division reached 176 by the end of the conflict, including numbers of young women working in a clerical capacity. Squeezed out of the State, War, and Navy Building in June 1917, Hunsaker and his people took over the top two floors of a building on the corner of Seventeenth Street and New York Avenue. Hunsaker remembered the Allies Inn on the ground floor as "a good place to get coffee" as well as a convenient location to meet with foreign officers who were in Washington to observe and consult with the Americans on aircraft design and production questions. The building also brought together Hunsaker's Bureau of Construction and Repair team and a group from the Office of the Chief of Naval Operations headed by Noble Irwin and John Towers, thus ensuring closer cooperation between the operational and the engineering sides of naval aviation. Other moves came in short order: to a temporary building at Fifth and B Streets in February 1918; and finally with the entire Bureau of Construction and Repair to the Main Navy Building at Nineteenth and B Streets (now Constitution Avenue).[39]

Along with expansion came reorganization. By the end of 1917 Hunsaker had established a separate Design Section, with two subsections—lighter-than-air and heavier-than-air. The lighter-than-air subsection incorporated Starr Truscott's design staff; the heavier-than-air subsection was formed under another civilian engineer, W. Starling Burgess, who had sold his Marblehead, Massachusetts, aircraft company to Curtiss before joining the navy as a reserve lieutenant commander. Hunsaker also set up the Production Section under another civilian, Carl A. Lohmann, who came to Washington from the B. F. Goodrich Company. The new Engineering Section (later renamed the Scientific Section) under Howard B. Luther, a professor of civil engineering at MIT and an expert in structures, had broad responsibilities for design and procurement. Among those Luther brought in to work with him in the Engineering Section were Charles J. McCarthy, also from MIT (and who went on to become chairman of the Chance Vought Corporation in the 1960s), and C. H. Congdon, who had previously worked as a research assistant at Curtiss. Luis de Florez, a 1912 MIT mechanical engineering graduate and later one of the navy's pioneers in the development of computers, came to the division in February 1918 with responsibility for the design and procurement of aircraft instruments.[40]

In 1918 Lohmann's Production Section split into two sections: Construction, which remained under Lohmann, and Supply, which became the responsibility of reserve Lt. Nelson Doubleday, scion of the publishing company Doubleday, Page. Hunsaker gave Doubleday complete authority over raw materials, priorities, and the shipment of aircraft, praising him after the war for doing his job "with a skill amounting to genius." Later Hunsaker assigned Louis M. Ream to work under Lohmann in the Construction Section, with primary responsibility for overseeing the flying boat programs. Hunsaker had first approached Ream, an executive with the American Steel and Wire Company, in August 1917, thinking that he was just the person to tackle the problems manufacturers were having with the wire cables used to actuate aircraft controls. After Ream joined the Aircraft Division in September, Hunsaker found that he had formidable administrative as well as technical skills, and the two became lifelong friends.[41]

Two other additions to Hunsaker's staff were John Jay Ide and Garland Fulton. A wealthy descendant of John Jay, the eighteenth-century diplomat and chief justice of the Supreme Court, Ide came from the editorial staff of the *Scientific American* with a fluency in French that helped make sense out of the technical documents often accompanying orders for aircraft material from France. Fulton, a lieutenant commander, joined the division as Hunsaker's executive officer in May 1918. He was a Naval Academy graduate, class of 1912. In 1914 he had entered the Construction Corps and took Hunsaker's aeronautical engineering course at MIT. Two years later he earned his master's degree in engineering. Hunsaker found Fulton to be a highly intelligent officer with an excellent understanding of aeronautical engineering. He also displayed the administrative competence to relieve Hunsaker of much of the division's routine business and to fill in for him when he was away from Washington.[42]

In the midst of aircraft production, Hunsaker and his staff undertook the design of a next-generation flying boat, significantly larger than the F-5-L. The impetus for the new airplane, which emerged as the famous NC-series of aircraft, came in late August 1917 from Taylor, who wanted a flying boat big enough to take off and land in virtually any sea state and with the range to cross the Atlantic—delivering itself, so to speak, thus freeing the navy's air arm from dependence on merchant shipping. Taylor met with Hunsaker and George Westervelt in his office to go over the idea of the big new flying boat. He wanted an airplane with transatlantic capability, a large capacity for bombs or depth charges, sufficient firepower to defend itself against other aircraft, and simple enough to be produced in numbers before the end of the war. Both Hunsaker

Lieutenant Commander Hunsaker (third from left), with Bureau of Construction and Repair officers, 1919. Seated in the middle is Rear Adm. David W. Taylor, chief of the Bureau of Construction and Repair. Hunsaker headed the bureau's Aircraft Division during and immediately after World War I. (Courtesy JCHP/MIT)

and Westervelt were skeptical about the concept, which they thought was too much for the immature American aircraft industry. Unconvinced by the two officers' reservations, Taylor ended the meeting with an order to his subordinates to "get busy and produce results."[43]

Hunsaker, Westervelt, and the Aircraft Division staff immediately began exploring various alternative designs, drawing on what they knew of current practice and adding a carefully measured dose of innovative ideas. During the preliminary stages of the project, Hunsaker and Westervelt added Glenn Curtiss and two of his engineers to create a high-powered navy-civilian design team. By the middle of September, they had two proposals: one, a biplane with five engines; the other, a similarly configured aircraft powered by three engines. Massive airplanes by contemporary standards, both featured a short hull from which outriggers extended aft to support the tail surfaces. An extrapolation of the smaller, single-engine Curtiss "flying lifeboat," the design offered the dual advantages of a compact and high-strength hull while elevating the tail group to a position isolated from the pounding of heavy seas. After reviewing the two proposals, Taylor leaned toward the five-engine design but finally opted for the smaller airplane, which was not nearly as complex and which presented a better chance of going into production without too much delay.[44]

Complexity in this case was a relative term, because the proposal still called for an airplane that pushed the state of the art. Hunsaker considered the application of theory and extensive wind tunnel tests crucial for the success of the project, because "an accident on the trial flights would set the project back many months besides risking the lives of the crew." The initial design called for a 140-foot (42.7-meter) wingspan, a wing area of 3,370 square feet (313 square meters), and a maximum takeoff weight of 20,000 pounds (9,072 kilograms). As the work progressed, however, the weight of the airplane quickly went up to 25,000 pounds (11,340 kilograms). At that critical point in the design process, Dick Richardson returned from Pensacola to take the job as inspector of naval aircraft at Curtiss Buffalo. A big, burly man, considered by many to have been the navy's first engineering test pilot, Richardson had only to glance at the data to conclude that excessive weight would seriously limit the flying boat's speed and endurance.[45]

Concerned that "the prestige of the Navy would be hurt by failure" of the airplane, Hunsaker and Westervelt went to Taylor with their problem, explaining to him that the original design was not going to work and recommending a smaller flying boat that would not have the range to cross the Atlantic nonstop. Much to the officers' relief, Taylor agreed with their assessment and told them to go ahead with a new design based on the assumption that the aircraft could fly as far as the Azores and refuel there before continuing on to France or Britain via Portugal. With Taylor on their side, Hunsaker and Westervelt convinced Curtiss that the flying boat project would need rethinking.[46]

The result of the redesign was an airplane of about 19,000 pounds (8,600 kilograms), with a wingspan of 106 feet (32.3 meters) and a wing area of 2,375 square feet (220.6 square meters). By the end of 1917 the anticipated increase in power of the V-1649 Liberty engine from 300 to 400 horsepower gave the engineers a little more leeway, allowing them to increase the flying boat's maximum takeoff weight to more than 21,000 pounds (9,525 kilograms). The wingspan and wing area also went up—to 126 feet (38.4 meters) and 2,380 square feet (221 square meters), respectively. In late November, when it became apparent to Hunsaker that to complete the preliminary design would involve practically all of the Aircraft Division's staff, the navy contracted with Curtiss to carry out the remainder of the detailed design work by Curtiss Engineering in Garden City under Westervelt's general supervision. Reflecting the cooperative nature of the project, the flying boat now received the designation NC, for Navy-Curtiss.[47]

Once the decision had been made to proceed with the smaller flying boat, Hunsaker and his staff in Washington had relatively little to do with the details of the NC design, most of the work being done by Curtiss's engineering department in Buffalo and by Curtiss Engineering in Garden City. But Hunsaker's office had responsibility for final approval of all aspects of the design, and Hunsaker himself participated in numerous conferences concerning the aircraft. Hunsaker knew, as well, that he had to keep an eye on the project, for he had previously expressed doubts about the competence of the Curtiss management at Garden City to administer major production contracts. To expedite NC manufacture, the navy paid for the construction of an assembly building adjacent to the existing facility at Garden City and let separate contracts for the major components of the NC boats.[48]

A cost-plus contract went to Curtiss Engineering for four of the airplanes. Curtiss was to build the first hull and subcontract to outside firms for the remaining three. The principal structural material was spruce, the idea being that, although metal construction offered advantages, an airplane skirting the edge of the design "envelope" ought to stay with a material whose properties were already well known. Delays set the NC-1's first test flight back to 4 October 1918. Considerable development work still lay ahead, particularly with the engine and propeller installations, and the remaining three boats were not completed until the spring of 1919, long after the cessation of hostilities in Europe.[49]

Hunsaker inspected the airplanes at Rockaway, Long Island, in late April, just before they took off for Newfoundland, the jumping-off point for the transatlantic flight. Taking full responsibility for the project, he admitted to Alice that he "will naturally get the blame if the boats fail to come up to expectations." He need not have worried. The NC-4 completed the Atlantic "hop" on 31 May, earning its place in history as the first aircraft to make the crossing. Despite all the problems, there was much for Hunsaker to be proud of in the NC program, for it marked the fruition of a large-scale design effort that relied heavily, though not exclusively, on aerodynamic theory and data derived from extensive wind tunnel and model basin experiments.[50]

Most engineers regarded World War I "not as a disaster for civilization but as a unique opportunity" to test their ideas of efficiency and cooperative government-business relationships.[51] If that were so, Hunsaker and his colleagues in Washington passed the exam. From virtually nothing, Hunsaker built an apparatus that brought together naval and civilian experts and integrated them into a team that swiftly and efficiently guided aircraft programs from their inception through design, production, and delivery overseas. Exhibiting the hallmarks of the heterogeneous engineer, Hunsaker combined intelligence,

The famous NC-4, in 1919 the first aircraft to fly across the Atlantic. Designed by Hunsaker in collaboration with George C. Westervelt and Holden C. Richardson, the big flying boat came from a directive in the Bureau of Construction and Repair for an airplane that could "ferry itself" across the ocean. (JCH 22, Courtesy MIT Museum)

perseverance, and technical knowledge with political and administrative deftness to make this triumph possible. New flying boat and airship designs originated in Hunsaker's office, which had responsibility also for the creation of a network of manufacturers and subcontractors for millions of dollars worth of aircraft and aviation equipment. The navy, in sharp contrast to the army, succeeded in getting significant numbers of airplanes to Europe and into operation, which was no mean accomplishment given the immensity of the task.

To a great extent, though, the Aircraft Division's decisions about numbers and types of aircraft to procure, although on the surface a model of efficiency and planning, were little more than responses to external pressure, much of it from Admiral Sims in London. Neither Hunsaker nor anyone else in the Bureau of Construction and Repair were able to come together with their counterparts on the operational side to effect a rational, long-term procurement plan intended to provide a balanced mix of aircraft for the fleet.

Lost in all the sound and fury in Washington was the shipboard scout, an airplane that, if it had been introduced during 1916–18, might have placed the

United States in a position of leadership in fleet aviation. Nor had there been much thinking until the fall of 1918 about the aircraft carrier, which the British found late in the war to have possibilities as an offensive weapons platform. To be fair, there was only so much Hunsaker could have done during the war, and he had to suppress his creative engineering talent and direct his energies toward accommodating present needs. Had the war gone on for another year, as most planners estimated it would, there would have been time to accomplish longer-term goals and to iron out administrative problems. But the war ended, and the sudden halt to the fighting presented Hunsaker and others with new problems, not the least of which being how to shape naval aviation so that it best met the needs of the fleet under greatly changed political, economic, and bureaucratic circumstances.

Peacetime Projects
4. and Reorganization

On Monday, 11 November 1918, the Armistice suddenly brought an end to the harsh realities of war and presented Americans and the world with the uncertainties of peace. The war had been a catastrophe of unprecedented scope, consuming millions of lives and the treasure of the world's wealthiest nations. At the same time, American naval aviation had come into its own. From next to nothing at the beginning of the war, the naval air arm had grown to more than 2,000 aircraft and more than 45,000 officers and men; forty-one air stations at home and overseas supported the navy's massive new air force. Manufacturers produced more than 1,100 single-engine and twin-engine flying boats, of which nearly 400 were delivered overseas. Most significant, American air patrols contributed to suppressing the submarine threat in the Atlantic.[1]

Jerome Hunsaker, now thirty-two years old, was pleased with the organizational and administrative effort he and his team had made in ensuring that American naval forces overseas had the numbers and types of airplanes needed to prosecute the war. Promoted to lieutenant commander in March 1918, and settled with his growing family in a rented house at 2333 Ashmead Place in northwest Washington, not far from Rock Creek Park, he hoped the end of the war would allow him to return to aeronautical design and engineering—and the immensely satisfying process of turning theory into the reality of airplanes and airships. He looked forward, also, to renewing his ties with the European aeronautical community, which had been broken since the onset of war in 1914.[2]

With his usual perceptiveness, Admiral Taylor knew that with the coming of peace it was urgent to glean aeronautical information from Britain and France before the inevitable veil of secrecy dropped over their military aviation programs. The end of the war meant, too, that the United States would have the opportunity to see what progress the Germans had made, especially with rigid airships. Admiral Sims concurred on the need to gather aviation intelligence in Europe, cabling Taylor on 8 November to say that he wanted Hunsaker in England "as soon as possible" and promising that he would provide assistance with the collection and compilation of technical data. Taylor decided to send Hunsaker to England with Goodyear's Paul Litchfield, Ralph Upson, and Clifford Slusser to scoop up as much intelligence as they could.[3]

Leaving the Aircraft Division in Garland Fulton's capable hands, Hunsaker took the train for New York on 15 November, arriving later in the day and putting up at the Hotel Astor. Although he was in the city only briefly, he found time to see his mother and her sisters Katharine and Docela before embarking on the White Star liner *Olympic* the next day. The atmosphere was almost festive: "The ship's company being small, and this sailing being the first since the Armistice, there has been a pronounced air of a holiday about everyone. We are running with everything lit up at night like a ferryboat," he wrote to Alice as the ship entered European waters. In addition to the navy contingent, the American passengers included Herbert Hoover and Edward N. Hurley, who were going to Europe to gather information on shipping in preparation for a postwar American food relief mission. Hunsaker had definite opinions about those he met on the ship: Hurley was "an ass"; Hoover "certainly not one"; and Litchfield and his group were so provincial that they could not see beyond the Akron city limits. Alice was probably glad to learn that her husband thought the army nurses on the ship looked "like somebody's maids dressed up . . . a lot of 'My Gawd' squawking. They probably look very beautiful to some sick chap, but in clear daylight look pretty crude. I am sure some of them are tough nuts who have got across to have a good time."[4]

The Americans arrived in Southampton on the morning of the twenty-third and immediately took the train to London to meet with Sims and fellow naval constructor, Comdr. Emory S. ("Jerry") Land, who along with Comdr. George Westervelt had been looking at aircraft production facilities and operating bases. To Hunsaker, war-weary London looked "grubby," with "huts" in the public squares, crowded buses, and streets still dark at night, but the hotel where he stayed near Hyde Park was comfortable. Sims briefed Hunsaker on his attachment to the aircraft section of the Allied Naval Armistice Commission, charged with investigating German naval aviation developments and with securing the surrender of German aircraft under the terms of the armistice.

Hunsaker's assistant was Ens. Ralph S. Barnaby, a youthful reserve officer and glider pilot, who had been in France and England since July helping to expedite the delivery of flying boats to operating units.[5]

On 5 December, the British and Americans crossed the North Sea to the port of Wilhelmshaven. Over the next three days, Hunsaker's party traveled by train to the airship operating stations at Wittmund, Ahlhorn, and Nordholz, returning to Wilhelmshaven each evening. Wracked by war, Germany gave every appearance of being a defeated power; many naval officers and men had deserted, there were no cattle in the pastures, and there were widespread food shortages and unemployment. Hunsaker recalled that the commission "went through the gesture of removing the firing pins from the German ships, and removing the ignition from the German bombers, in case some crazy Germans wanted to do a last dying effort to bomb somebody." In Kiel, the big German naval base on the Baltic, Hunsaker made pencil sketches of the Dornier Rs.IV, a giant four-engine, all-metal flying boat, with detailed comments on its duralumin construction. He also visited Tondern, where British aircraft from the carrier *Furious* had attacked and destroyed two airship hangars, before returning to England on 16 December.[6]

During a lonely Christmas week in London, Hunsaker busied himself with the idea of acquiring a large rigid airship. He reported to Taylor on 21 December that the United States had two options for quickly acquiring a rigid airship: having the British build one similar to their R.34 or having the Germans build one of the most modern type under the terms of the peace agreement. He also stressed the urgency of obtaining confidential information from the British. At the end of January, he wrote to Assistant Secretary of the Navy Franklin D. Roosevelt that "just now, the British policy is to help us" in rigid airship development, and that any hesitation or delay in getting a program going would mean "we may have to go it alone." On his own, Hunsaker befriended Comdr. Charles I. R. Campbell, a naval constructor in charge of British rigid airship development, and spent many hours with him discussing specific problems associated with airship design and construction. From the French he obtained a set of plans for the German airship L.49, a "trunkful of drawings," he said, which provided the most reliable firsthand data the Americans were able to secure during their mission.[7]

Concluding his work for the Allied Naval Armistice Commission, Hunsaker spent a month preparing more than ninety reports on British, French, and Italian aviation developments. He covered a variety of subjects, including the details of British airship construction, mooring equipment, corrosion prevention, strength testing and equipment, bases, wartime operations, duralumin, and nonrigid airship envelope construction and materials. He also analyzed big, multiengine airplanes. Caproni, for example, was designing three-engine air-planes and looking toward passenger service in the Mediterranean. In Britain, Handley Page had a four-engine airplane "fully developed," and John Cyril Porte, who had developed the large F-type flying boats during the war, was working on a five-engine triplane capable of flying the Atlantic. Ralph Barnaby found that Hunsaker had a clever technique for prying information from people: When he ran into someone who was hesitant about divulging much about an engineering problem, he immediately took the offensive, declaring that it was obvious the person knew nothing about the subject and that the Americans had come up with a better solution. Usually that was all he needed to get the reluctant expert to reveal everything he had been holding back.[8]

His reports filed, Hunsaker sailed from Liverpool on the Cunard liner *Aquitania* on 20 February 1919, arriving in New York on the twenty-eighth. Among *Aquitania*'s passengers was Brig. Gen. William ("Billy") Mitchell. Fresh from service as commander of the American Air Service in France, Mitchell expounded to all those within earshot about the preeminence of air power, the war-winning potential of land-based airplanes in the strategic bombing role, and the need for merging army and navy aviation into a unified air force. Hunsaker listened intently, for he knew that Mitchell's rhetoric spelled danger for the navy in general and for naval aviation in particular. Mitchell and his cohorts, Hunsaker recalled, were "fully prepared with evidence, plans, data, propaganda posters, and articles to 'break things wide open'" as soon as they arrived back in the country. Hunsaker took an immediate personal dislike to Mitchell, whom he regarded as a masterful "politician in uniform," "charming" in some ways but with a certain "asinine quality." Regardless, Mitchell had to be taken seriously, and Hunsaker resolved that he had to alert his fellow officers to the probable menace.[9]

Back at his desk in Washington on 1 March, Hunsaker went to work digesting the information he had acquired and preparing for hearings by the General Board of the Navy on postwar naval aviation. The navy's senior policy advisory body, the board placed top priority on the acquisition of a high-performance shipboard airplane. Fleet exercises helped build a consensus during 1919 that aircraft were needed as spotters, scouts, and fighters with the fleet. Equipped with floats, the airplanes could be catapulted from battleships and cruisers and recovered alongside. But only airplanes with wheeled undercarriages could achieve maximum performance, and to operate them the navy needed an aircraft carrier. After extended discussion and analysis, the board in the spring of 1919 recommended the conversion of the collier *Jupiter* into the navy's first carrier, to be named the *Langley*. Hunsaker's responsibility was designing an airplane to operate from the carrier as well as from existing battleships and cruisers.[10]

From his experience with the Speed Scout, Hunsaker knew that a shipboard fighter was a tough design proposition. The major obstacles he and the Aircraft Division encountered centered on the sometimes conflicting requirements for high performance, compact size, and extended range. Despite Hunsaker's warnings that he knew of no manufacturer who could supply an airplane meeting the navy's requirements, the General Board went ahead with a request for a fast, high-powered, single-seat airplane mounting two machine guns constructed to withstand the rigors of ship operations.[11]

Only two manufacturers submitted designs for the shipboard fighter, and neither came close to meeting the navy's performance criteria. Admiral Taylor then turned the whole project over to Hunsaker in the spring of 1921, with the understanding that once the Aircraft Division came up with a satisfactory design, it would be handed over to one of the manufacturers for production. The result was the TS-1, considered by many historians and engineers to be an aeronautical tour de force. Among the airplane's outstanding features were a fuel tank incorporated into the lower wing's center section and the omission of the drag-inducing wire braces that characterized most biplanes of the era. With a maximum takeoff weight of 2,133 pounds (968 kilograms) and a wingspan of only twenty-five feet (7.6 meters), the airplane was small enough to be handled easily onboard ship and was convertible between conventional landing gear and twin floats. A 200-horsepower Lawrance air-cooled radial engine provided a top speed of more than 120 miles (193 kilometers) per hour. Curtiss and the Naval Aircraft Factory produced the airplane, which after trials in the summer of 1922 went to the fleet, where it put in exemplary service.[12]

Among Hunsaker's and the General Board's other considerations in early 1919 was the rigid airship. Armed with a battery of maps, charts, and graphs, Hunsaker spent two days in April detailing German airship operations over the North Sea and providing the board with information on basing arrangements. For the navy's planned airship hangar and base at Lakehurst in the pine barrens of New Jersey, Hunsaker stressed the importance of constructing the hangar large enough to accommodate present and future craft and having enough acreage to allow safe ground handling. An adequate labor supply was essential, especially if the airship construction facility were located, like Lakehurst, outside a major metropolitan area. To avoid this problem, Hunsaker suggested carrying out most of the major fabrication at the Naval Aircraft Factory and transporting the components to Lakehurst for final assembly. Hunsaker advised that the best approach would be to order an airship from the British on the as-

The TS-1 on floats. Another Hunsaker design, the airplane with wheels proved to be an excellent shipboard fighter after it went into service in 1922. (18-WP-29850, Courtesy National Archives)

sumption that there would be improvements in its design and construction comparable to those in the latest German versions.[13]

The scale of the American airship program also interested the General Board. Hunsaker told the board later in the month that during the war the Germans' superb industrial organization had enabled them to sustain airship production, making good their losses while steadily improving the airships' design and performance. With advanced planning and sufficient funding, there was no reason why the United States could not do the same. Hunsaker contemplated building two airships annually, starting in 1921 and eventually building up to a fleet of ten by 1926. He explained that "ten is not a mystic number."

> I have chosen that because it appears reasonable to build two a year at one plant. Of those ten we would expect to lose—based on German experience—four ships . . . so that in 1926 we should reasonably expect to have at least six . . . and thereafter if we should build two ships per year and junk two or lose two we maintain the six.

Insisting that "we have to have a program," Hunsaker urged the General Board to recommend as soon as possible the acquisition of an airship from abroad, to secure enough funds to start the design of an airship in the United States, and to move quickly with construction of the operating base and airship hangar at Lakehurst.[14]

Given the postwar budget restrictions that the navy would soon face, an airship program of the scope Hunsaker envisioned in the spring of 1919 was hopelessly optimistic. On the other hand, Hunsaker was realistic about operational losses and believed that if the navy and the nation were serious about airships a long-term program with a major financial and industrial commitment was essential. Building just one or two of these expensive and complex aircraft made no sense, involving as it did a huge financial outlay without providing enough units to permit sustained operations. Nor would the nation acquire the infrastructure needed to support a new industry with potential commercial applications. Hunsaker presented a convincing case, for the General Board acted with surprising alacrity. In early May it recommended, and Secretary of the Navy Daniels subsequently approved, the procurement of two airships—one to be built in Britain, the other in the United States—as well as the establishment of a base and all necessary support facilities. On 11 July Congress earmarked $6 million of the $25 million fiscal year 1920 naval aviation budget to implement the program, half of which was for the airships themselves and the other half for the hangar and operating complex at Lakehurst.[15]

In connection with the navy's airship program, Hunsaker was detailed to Mineola, Long Island, where the British R.34 touched down on 6 July 1919 following a grueling transatlantic crossing. Before the airship's return to Britain on 10 July, completing the first two-way transatlantic flight, Hunsaker once more studied the details of its construction and discussed airship operations with its officers and American navy Lt. Comdr. Zachary Lansdowne, who had been aboard as an observer. The success of the R.34 flight and the intense publicity surrounding it was a shot in the arm for American airship advocates, timed perfectly with the congressional airship appropriation.[16]

Negotiations to acquire an airship from the British Admiralty, which retained responsibility for airships after the creation of the Royal Air Force in 1918, opened old wounds between the aviators in Naval Operations and the engineers in the Bureau of Construction and Repair. Hunsaker told Comdr. Lewis B. McBride, a Construction Corps officer assigned to the attaché's office in London, that there was "friction" between the two. The aviators, he said, "seem to have the idea that it is their pigeon and they will be advised by the other Bureaus when, if, and as necessary." Admiral Taylor saw it differently. To him the matter was "a straight ship building job," with McBride as the bureau's representative having full authority to make all arrangements with the British for buying the airship. As a compromise, the bureau handled the details of the purchase and Naval Operations negotiated an understanding on the training of the airship's crew. It was at best a clumsy arrangement, workable only through the efforts of McBride and others in the attaché's office in London.[17]

The British and Americans finalized the purchase agreement for the airship, designated the R.38, on 5 December 1919. Commander Campbell had designed the 2.9-million-cubic-foot (82,120-cubic-meter) airship in the late summer of 1918, basing his plans on data garnered from the wrecked German Zeppelin L.70, one of the most up-to-date craft in the German inventory. The cost to the United States was not to exceed $2 million, with delivery to take place by the fall of 1920. Because the navy had contracted with the British Air Ministry (which had assumed responsibility for airship development from the British Admiralty) rather than a private manufacturer, it was never clear what authority the American inspection officer had at the airship base in Cardington. Hunsaker was not entirely pleased with the situation, but at least he had a good relationship with Campbell, who had worked as a consultant in the Aircraft Division in December, helping draw up plans for the American-built airship. Little did he know at the time that much of the R.38 design was based on sheer guesswork rather than sound aerodynamic research.[18]

In the meantime, Hunsaker and the Aircraft Division's lighter-than-air design staff were hard at work on the home-brewed airship. Designated the FA-1 (for Fleet Airship 1), the craft was to be fabricated at the Naval Aircraft Factory and the components shipped by rail to Lakehurst for final assembly. Safe, inert helium, rather than flammable hydrogen, was the lifting gas. As project engineer, Starr Truscott used the design of the wartime Zeppelin L.49 as a starting point, drawing heavily on information obtained from the French on the German airship L.72 as well as from the British work on R.38. Because this was a learning experience, Hunsaker wanted to stay within the boundaries of known airship design. Early in the project, he told a friend in the British Air Ministry that "we don't want to be too radical, and we feel rather humble about the skill with which we will attack this problem. We're sure to make mistakes." One of the mistakes Hunsaker did not want to make was in the strength of the airship. He enlisted the help of Charles Burgess and the distinguished MIT naval engineer William Hovgaard to generate what at the time were probably the most complete airship stress calculations ever compiled. Congressional funding delays held up the airship's design, which was not finished until October 1920, pushing the anticipated start of construction back until 1921.[19]

Hunsaker was all too aware that the acquisition of the British airship and the training of its crew had been complicated by the divided responsibility for naval aviation and believed that the time had come for fundamental administrative reform. In the late winter and early spring of 1919, the General Board turned to Hunsaker and other technical officers to probe the necessity for organizational change and the legitimacy of Billy Mitchell's ideas about the superiority of land-based aircraft and a unified air force. Hunsaker testified before the board on 10 March that for torpedo and bombing aircraft to damage or destroy ships they needed to be operated from ships at sea. "The Germans had torpedo planes and abandoned them because they had no airplane carriers," he said. There was nothing to indicate that land-based air power of the sort Mitchell wanted would have any success in a future naval conflict. Hunsaker urged the board to call on Mitchell, because he "has the idea that the Navy should be absorbed by the Army and I would like to hear him expand his views on that."[20]

The General Board followed up on Hunsaker's suggestion and invited Mitchell to appear on 3 April. True to form, Mitchell expounded on a wide range of topics, telling the board that the fighter aircraft the navy was experimenting with for fleet use would be "shot down as fast as they go up" and that army land planes could handle coastal patrols more effectively than navy flying boats. Most ominously, he argued that the naval air arm should be consolidated into a single, independent air force, subsumed under a combined Ministry of Defense.[21]

Mitchell's testimony threw down the gauntlet in what was to become a long and bitter quarrel over aviation policy, pitting naval officers against the advocates of land-based air power in a high-stakes, winner-take-all game. Initially, Mitchell's challenge drove Hunsaker to the conclusion that something had to be done about the administration of naval aviation. He saw organizational change as a "logical" result of technological change, flowing from "the development of the art and its application in the Naval Service." Technical officers, in particular, believed that the divided responsibilities between the Office of the Chief of Naval Operations and the bureaus had caused inefficiencies, but they had put up with the situation as long as there was plenty of money and a general understanding of the need to pull together in a common cause. Hunsaker later wrote that "in general, the organization was bad, but the results [were] good." After the war, the setup proved unacceptable when traditional bureaucratic rivalries resurfaced as the navy underwent a drastic reduction in personnel and equipment.[22]

In testimony before the General Board on 9 May 1919, Hunsaker couched his views on the reorganization of naval aviation in terms of the design, development, and procurement of aircraft. He explained that the current organization worked under the assumption that the Office of the Chief of Naval Operations was the "customer" and the material bureaus were the "manufacturers and merchants who find out the needs of the customer" and tried to fill them. Yet without clear-cut lines of authority, "the detail of doing business between a lot of bureaus means that some one bureau must be responsible for digging out the material from the others." The Bureau of Construction and Repair had done this on a purely ad hoc basis. "It works well now," he maintained, "but if you change the personnel and someone comes in who does not want to work with the rest, it might go all to pieces." During the war, the bureau had been in the position of "doing things we had no business to do" and that additional responsibilities had been "forced on us by the lack of policy and far sighted planning from top side." With the NC boats, for example, the bureau and not the operations people had identified the problems of dispatching aircraft overseas, solicited information from the operating personnel, and finally determined the requirement for a "self-ferrying" flying boat.[23]

Hunsaker thought that giving more authority to operations would do little to solve the problem, because Capt. Thomas T. Craven, who had succeeded Noble Irwin as director of naval aviation, had no intention of his office taking on aircraft design and development, and the additional responsibility would do nothing to reduce interbureau frictions. Nor was the solution a separate aviation bureau, as some had suggested. Asked specifically what he thought should be changed, Hunsaker said that it might be best "to leave the different bureaus cognizance of the materials they are concerned with but give to one bureau the problem of getting together and installing the material." It was clear that he preferred the one bureau to be the Bureau of Construction and Repair. He added that an experienced engineering officer should be assigned as an aide to the director of naval aviation to help promulgate operational requirements and advise on aircraft design and development.[24]

Following up on some of the ideas he had presented before the General Board, Hunsaker consolidated his thoughts about the organization of naval aviation in a memo to Rear Admiral Taylor on 18 July. He continued to emphasize conservative reform, insisting on the segregation of design, development, and procurement from the operational mission of naval aviation. Problems had arisen in part because the director of naval aviation, in the Office of the Chief of Naval Operations, "in attempting to coordinate, accelerate and stimulate" naval aviation, had "added confusion and friction unnecessarily by duplication of Bureaus' work and by encroachment upon proper Bureau functions," generating an atmosphere rife with "petty irritation and annoyance." Rather than a separate bureau, he wanted to see all policy and planning responsibilities left to the director of naval aviation and a chief aeronautical engineer within the Bureau of Construction and Repair charged with "full responsibility" for the material side of aviation. An experienced officer from the bureau could assist the director of naval aviation as necessary.[25] Hunsaker's proposal was largely self-serving, because it was apparent to Taylor and others that Hunsaker had himself in mind as the man responsible for aviation within the Bureau of Construction and Repair.

Admiral Benson's action in the summer of 1919 to abolish Craven's staff and relegate the functions of the Office of the Director of Naval Aviation to the Naval Operations' Planning and Material divisions was the last straw, convincing Craven that the only way to redress the situation was to create a separate bureau with broad-ranging authority over naval aviation. With the backing of the new chief of naval operations, Adm. Robert E. Coontz, Craven persuaded Long Island Republican Congressman Frederick C. Hicks to go ahead with a bill to create a bureau of aeronautics. But before that measure made any headway, Congressman Fiorello H. La Guardia, a New York Republican, convened hearings of a subcommittee of the House Military Affairs Committee on the question of a unified air service. A World War I army aviator, friend of Billy Mitchell, and future mayor of New York City, La Guardia pointed to waste and inefficiency in military air operations and wartime aircraft procurement as reasons for bringing army and navy aviation together under one umbrella.[26]

Fearing that La Guardia's subcommittee was hearing only one side of the story, Craven counterattacked in force, bringing Hunsaker and six other naval officers with him to testify on 20 December. Carefully choosing his words, Hunsaker declared that the functions of navy and army aviation were distinct and that the ultimate solution to the postwar problems in military and civil aviation was "more money." Hunsaker chose that moment to take the offensive, asking La Guardia if he would "give the new department more than the sum total of the appropriations to the separate departments." La Guardia responded, "No, we would give it less." To Hunsaker that was a false economy. "If you were to give it the same or less," he said, "I believe you would have less aviation in the air than if your appropriations were made direct to the departments which already have the administrative machinery to handle their aviation without spending one cent of aviation money to run it." The system could "not be improved by beginning over again," and it made no economic or administrative sense to graft a new bureaucracy onto the ones already in existence in the war and navy departments.[27]

Away from the congressional spotlight, Hunsaker summed up his feelings on the issue of unification in a letter to his friend from MIT days, Comdr. Ralph T. Hanson. "Naval Aviation," Hunsaker wrote, "just now is in a bad way because of the agitation in and out of Congress for the establishment of an independent Air Service, co-equal with the Army and Navy." He considered the creation of a separate air force a "high crime" that would primarily serve as a means for the "boy generals" and "boy colonels" to preserve their rank and salvage their careers by building an empire of their own. Partisan politics also played a role. That fall Republican Senator Harry S. New of Indiana had introduced legislation to bring military and civilian aviation together in a separate air department. The New bill implicitly criticized the administration for the way it had handled aviation and aircraft procurement during the war and as a result had the support of many Republicans, who were looking ahead to unseating the Democrats in the 1920 elections.[28]

The New bill did not pass, but the support it received on Capitol Hill convinced most naval officers that the navy had to face the reorganization issue squarely or risk losing control over the service's air arm. In February 1920 the National Advisory Committee for Aeronautics drafted a measure that recommended against an independent air force and called for the creation of aeronautics bureaus within the army, navy, Post Office, and Department of Commerce. Still worried that technical officers would lose their influence in a bureau with sweeping authority for aviation, Hunsaker put together an alternative proposal that incorporated an Aviation Corps to deal with operations and personnel and a separate bureau primarily responsible for aviation material. Taylor approved Hunsaker's draft and included its main points in a 26 February memorandum to Secretary Daniels, who agreed to the inclusion of both a corps and a bureau in draft legislation. Congressman Hicks used the navy's draft as the basis for another reorganization bill, only to have the measure encounter opposition from the powerful Bureau of Navigation, which opposed an aviation corps. Rather than press the issue and exacerbate the divisions within the service, Hicks let the measure expire in committee and waited for a more propitious moment to reintroduce the bill.[29]

At this juncture in the spring of 1920, Hunsaker received an invitation to deliver the prestigious Wilbur Wright Memorial address before the Royal Aeronautical Society in London. Hunsaker took the speech as an opportunity to get away from Washington politics, see how the R.38 was coming along, and to assess European aeronautical work at the request of Joseph S. Ames, chairman of the NACA Executive Committee. Finding that all the big, fast transatlantic liners were booked, Hunsaker had to jam himself aboard the army transport *Mercury* sailing from New York on 19 May 1920. After finally arriving in Britain, Hunsaker drove up to Cardington on 9 June with Jerry Land to inspect the R.38. In a handwritten letter to Admiral Taylor, Hunsaker reported that the airship was "progressing well," even though the work had been held up while alterations were made to strengthen the airship's main frames. Alarm bells should have gone off in Hunsaker's head, because he knew that the British had done an inadequate job with their strength calculations, but he assured Taylor that the problem was "not serious."[30]

Hunsaker spoke to the Royal Aeronautical Society in Westminster's Central Hall on 22 June, with the Duke of York (the future King George VI) presiding. His presentation, "Naval Architecture in Aeronautics," was a classic treatment of the subject, comparing the work of the aeronautical engineer and the naval architect. Hunsaker concluded that although they had much in common the naval architect and the aeronautical engineer fundamentally differed in the way they approached their professions. "The Naval Architect is a craftsman with both artistic and scientific traditions," Hunsaker stated, "and the art he practices has a technique perfected by the experience of generations." With hundreds of years of experience, the ship designer relied mostly on empirical methodology, his main task being to sift through a vast store of information and identify that which was most useful. The aeronautical engineer could learn much from the naval architect in regard to weight control, stability, model testing, and statistical analysis. That "common ground" diverged, however, because there were many areas in which the aeronautical engineer lacked experience and data and had to rely on engineering science, which rested firmly on theory and applied mathematics.[31]

His responsibilities in London over, Hunsaker headed for Germany in early July. He found "the trains are dirty, in bad repair, and often infested with fleas." The same was true of German aviation, which was at a "very low ebb," constrained by the Allied Control Commission from military construction and by the country's feeble economy from commercial development. One of his top priorities was to visit the University of Göttingen and renew his ties with Ludwig Prandtl. He had the utmost respect for Prandtl's work, which he considered a model of engineering science. It was "of a class by itself because [it was] always done with a system of models forming a connecting series, and the results are analyzed and simplified by some working hypothesis." Hunsaker arrived in Göttingen during the school's commencement, when a general "holiday spirit" prevailed. Prandtl was eager to "show off" his lab and to share his latest aerodynamic theories with the Americans. Hunsaker negotiated an agreement whereby Prandtl was paid by the NACA to prepare a report, "Applications of Modern Hydrodynamics to Aeronautics," which appeared as an NACA technical report in 1921.[32]

Possibly the most important result of Hunsaker's contacts at Göttingen was the emigration to the United States of one of Prandtl's most outstanding students, the thirty-year-old Max M. Munk. In his doctoral work at Göttingen, Munk had collaborated with Prandtl in developing sophisticated mathematical models to study the flow of air over a wing, with the goal of minimizing the induced drag of airfoils. Shortly after the end of the war, Prandtl suggested to Hunsaker that he might help Munk find employment with the NACA. Hunsaker then sounded out Ames, who was receptive to the idea of bringing such a promising young aerodynamicist into the organization.[33]

Munk met Hunsaker and discussed with him his research interests, particularly some of his ideas about the Reynolds number in calculating wind tunnel data. In experiments with water moving around objects in pipes, Osborne Reynolds, a nineteenth-century Irish-born mechanical engineer, had determined that the flow over a scale model and a full-size object was identical provided the researcher adhered to a mathematically derived ratio of density, velocity, viscosity, and diameter of the object. That ratio—the Reynolds number—provided a correlation between the results of wind tunnel testing with models and the performance of full-scale aircraft; the higher the Reynolds number, the more accurate wind tunnel figures were in predicting the real performance of an airplane or airship. Hunsaker knew that often there were discrepancies in wind tunnel data when small models were used to generate information on large aerodynamic structures such as airships, and he was wary of the results of tests on small airship models because of the low Reynolds numbers involved.[34]

Impressed with Munk's knowledge and abilities, Hunsaker recommended that the NACA extend Munk an offer to work at the Langley Aeronautical Laboratory in Hampton, Virginia. Munk accepted without hesitation. (Later, he would thank Hunsaker profusely for giving him the opportunity to continue his aerodynamic work in the United States.) At Langley, Munk transformed his ideas into hardware. He designed Langley's variable-density wind tunnel, a brilliantly innovative apparatus in which pressurized air flowing over the model effectively multiplied the Reynolds number and greatly increased the accuracy of the resulting data.[35]

Hunsaker went on to visit German aircraft manufacturing facilities. At the Junkers works at Dessau and at Staaken, outside Berlin, where the Zeppelin company had built its "giant" R planes during the war, he saw what at that time was the world's most advanced metal aircraft construction. Moving on to the Zeppelin works at Friedrichshafen, he met two other naval officers, Comdr. Ralph D. Weyerbacher, who was touring German manufacturing plants to gather information for the FA-1 airship, and Jerry Land, who was there to sound out the Germans on the possibility of buying an airship. The director of Luftschiffbau Zeppelin, Alfred Colsman, warmly greeted the Americans and allowed them to inspect the small airships *Bodensee* and *Nordstern,* but as usual he would not reveal any information on their design and construction. Hunsaker told Admiral Taylor that the Americans were permitted "nothing more than a chance to look around and to be supplied with unlimited beer." The Deutsche Versuchsanstalt fur Flugwesen Luftfahrt at the Adlershof near Berlin, which he and Zahm had found before the war to be one of the world's best flight test installations, had been virtually shut down by the Allied Control Commission, and large quantities of the equipment had either been destroyed or confiscated. Nevertheless, through his friend August von Parseval, Hunsaker arranged to buy equipment for recording information from high-speed test flights.[36]

Soon after Hunsaker returned home—on the *Aquitania* on 21 August—he became deeply involved in the design of a torpedo bomber, considered another priority in the postwar years. Donald Douglas, his former student and research assistant at MIT, collaborated on the project. In the early spring of 1920, Douglas had moved to southern California, where he planned to get into aircraft manufacturing. He wrote to Hunsaker soon after settling in with his family near Los Angeles that he hoped "his friends in the East will not forget me, and believe that some day I shall be in a position here to come to you for any business that you may be contemplating farming out to the aircraft trade." With local financial help he organized his own company and cast about for business. Hunsaker had just the job for him: a twin-float seaplane large enough to haul a full-sized torpedo. Douglas returned to Washington, where he and Hunsaker worked out the design, which in 1921 emerged as the Douglas DT, the company's first major contract and the beginning of a long association between the navy and the California firm.[37]

Another manufacturer with whom Hunsaker had close personal ties in the early 1920s was pioneer aviator Glenn Curtiss. The two had met for the first time in 1912, when Hunsaker was in Hammondsport, New York, to tour the Curtiss shop and report on floatplanes and flying boats. Hunsaker recalled that Curtiss had been a "genial and considerate host," and he had extended his visit so that he and Curtiss could shoot pheasants among the vineyards along the banks of Lake Keuka. This professional acquaintance grew into friendship during the war years, when Hunsaker traveled to Hammondsport and Buffalo and spent time in Garden City during the development and construction of the NC boats. In the fall of 1920 Curtiss invited Hunsaker to join him for Thanksgiving dinner at his home in Garden City and sent him a hunting cap and some waterproof boot dressing. Unable to be there for the meal due to his responsibilities at the Pulitzer Trophy air races at nearby Mitchel Field, Hunsaker thanked Curtiss for the gifts and said that the hat and boot dressing would "save my life" during hunting trips.[38] Hunsaker viewed such personal, professional, and corporate ties as building blocks in a larger cooperative effort to advance the development of aeronautics.

After a hiatus of more than eight months, the naval aviation organization issue resurfaced, precipitated in January 1921 when Billy Mitchell charged in Congress that the navy had covered up the results of a bombing experiment on the old battleship *Indiana,* because the service was unwilling to admit that aerial weapons could destroy "any ship in existence today." Hicks, in the meantime, had reintroduced his bill, including a bureau of aeronautics and dropping the separate aviation corps, but the legislative session ran out before the bill reached the floor of the House. As the political drama played itself out on Capitol Hill, Capt. William A. Moffett succeeded Craven as director of naval aviation on 7 March 1921. Hunsaker liked the choice of Moffett, a battleship skipper, knowing that he would stand up to Mitchell and that because he enjoyed the complete confidence of the traditional "battleship admirals" was not likely to allow the operational faction to dominate the new organization.[39]

By now Hunsaker was resigned to a separate bureau as the "best thing for the Navy," although he still worried that if aviators assumed complete authority there would be no room for technical officers. In January 1921 he wrote to Comdr. Lewis M. Maxfield, who had been assigned to the naval detachment slated to man the R.38, that "it appears to be the idea of the particular crowd in control just now that they don't want any constructors in this new Bureau, but would prefer to develop their own technical people from ex-fliers." He told Maxfield that he would not be at all surprised if after the reorganization he would be completely divorced from aeronautics and would have to go back to working on ships. On 11 April Hicks submitted a new bill in the House, followed two days later by a similar measure in the Senate. On 12 July 1921 both houses passed the bills as parts of the Naval Appropriations Act for the next fiscal year, and President Warren Harding signed the measure into law the following day.[40]

The creation of the Bureau of Aeronautics was an important victory for naval aviation in the protracted war with Billy Mitchell over a unified air force. Mitchell was down but not yet out, and in the early summer of 1921 he scored his greatest propaganda coup during the famous bombing experiments off the Virginia Capes. In an effort to gather more data about the vulnerability of modern warships to aerial bombs, the navy organized an extensive series of experiments using captured German vessels as targets. No one wanted to miss the climax of the experiments on 21 July, when Mitchell's army bombers were to join navy aircraft in attacking the dreadnought *Ostfriesland.* Hunsaker, along with scores of other officers and civilians, left Washington on the nineteenth aboard the transport *Henderson* for the day-long voyage to the test site off the mouth of Chesapeake Bay. The carefully choreographed procedures called for navy aircraft, then army bombers, to deliver their ordnance, after which an inspection team would board the ship to assess the damage. On the twenty-first, the army fliers appeared at the scene earlier than anticipated and dropped their bombs first. Mitchell's pleas that the mix-up was inadvertent seemed credible until the next day, when his aircraft again arrived over the target before the navy airplanes and sank the *Ostfriesland* with 1,000-pound (454-kilogram) and 2,000-pound (908-kilogram) bombs. It all happened so fast that navy inspectors had no chance to conduct damage assessment, and whatever information that might have been obtained went to the bottom of the Atlantic.[41]

Reflecting on the bombing tests years later, Hunsaker concluded that "Mitchell was pretty bad" and that most naval officers did not believe the premature appearance of the army bombers was accidental. At the time, though, Hunsaker hoped the controversy would subside while the Bureau of Aeronautics established itself. He assured an army friend, the pioneer aviator Benjamin Foulois, that the navy wanted "friendly relations" with the army and that Foulois should tell Mitchell that "the Navy is not a crowd of pirates that are trying to put him in jail." Hunsaker's message never got through, because Mitchell used the bombing tests as a springboard for another, even more intense, round in his propaganda campaign, emphasizing the superiority of the bomber to the battleship and renewing his agitation for a separate air force.[42]

If Mitchell still represented a threat to naval aviation in general, the loss of the R.38 (now with the American designation ZR-2) was a potentially crippling blow to the navy's rigid airship program. Despite earlier indications that the airship might have structural deficiencies and numerous delays in her construction, everyone seemed to breathe easier when the airship began its test flights in June 1921 and they looked forward to her transatlantic delivery later that summer. Then, on 24 August, on only its fourth flight, with Maxfield and sixteen other American officers and men aboard with the British crew, the ZR-2 suffered a catastrophic structural failure, and its hydrogen lifting gas ignited during high-speed maneuvers over the Humber estuary. Forty-four died, including Maxfield and C. I. R. Campbell; only one of the Americans survived. Hunsaker was devastated—Maxfield and Campbell were his friends, and he knew many of the others who had been killed.[43]

Aside from Hunsaker's distress about the enormous personal loss caused by the ZR-2 disaster, he also fretted about the effects the accident would have on the navy's airship program. Above all, he wanted to know what had caused the crash so that everything could be done in the design and construction of the domestic airship (now designated the ZR-1) to ensure its airworthiness. In the fall of 1921, Charles Burgess sat in on the investigation conducted by the British Aeronautical Research Committee, which determined that the ZR-2's design was based on inadequate aerodynamic information and that fundamental structural weaknesses led to the accident. Burgess pursued his own contacts in Eng-

land, reporting to Hunsaker in late November that the British had "entirely neglected" the most elementary aerodynamic considerations in the design and construction of the airship.[44]

Also much on Hunsaker's mind in the summer of 1921 was the organization
of the new Bureau of Aeronautics. On 1 August, Moffett, elevated to rear admiral and chief of the new bureau, circulated a draft organizational plan calling
for four divisions within the bureau—Plans, Administration, Material, and
Flight. The assistant bureau chief also served as head of the Plans Division.
Two days after receiving the draft circular, Hunsaker sent Moffett a memo ex-
pressing reservations that the Plans Division seemed to have more status than
the other divisions and that engineering officers from the Bureau of Construction and Repair were likely to be excluded from the new bureau. Hunsaker fa-
vored a clear distinction between the responsibilities of officers in the Plans Division and those in the Material Division. It was best, he told Moffett, to have
the Plans Division staffed mostly by aviators drawn from the Office of the
Chief of Naval Operations, who would develop broad guidelines for action,
while engineers from the Bureau of Construction and Repair would be in
charge of the Material Division and have wide latitude for developing the aircraft needed to carry out the mission. Technical officers in the Material Division also had to have direct access to the bureau chief. Only they knew whether
or not a given innovation warranted the bureau's attention and money, and they
should not have to go through the Plans Division every time they came up with
a new idea.[45]

In its final form, Moffett's organizational blueprint followed Hunsaker's suggestion to separate the assistant chief from the head of the Plans Division. Nor
did Hunsaker have cause for anxiety about the status of technical people in the
new bureau. When the formal transfer occurred on 1 September, fifty-one officers and civilians moved over from the old Aircraft Division of the Bureau of
Construction and Repair, constituting 55 percent of the bureau's total personnel. The organization was familiar to anyone from the BUC&R: the Material
Division comprised six sections, with the Design Section responsible for the
specifications, design, and procurement of aircraft, power plants, and ancillary
equipment. Design, in turn, was divided into lighter-than-air, heavier-than-air,
and engine subsections. The Material Division's other sections were Maintenance, Lighter-than-Air, Engine, Scientific, and Technical Information. It was
clear to everyone that technical officers would have much to say about how
Moffett and the new bureau went about their business.[46]

Largely as a result of his administrative experience as head of the old BUC&R
Aircraft Division, Hunsaker was the prime candidate to head the Material Division. On reflection, he thought better of the idea. In December 1920 he had

written to Jerry Land, who was still in London, to ask him if he might want to take over his job before the new bureau was approved. A bright and highly capable officer, Land had graduated from the Naval Academy in 1902 and had done postgraduate work at MIT before joining the Construction Corps. Hunsaker praised Land for his "administrative skill and savoir faire" and guessed that he would be acceptable to the aviators and operations people. In contrast, Hunsaker judged that he, himself, carried too much political baggage and that some of his fellow officers believed he had been in the position so long that he "thinks he owns aviation." Land's appointment would be a "morale builder" for the technical officers in the BUC&R, which was essential if they were not to be treated as second-class citizens or regarded as "servants to aviators." Land protested that Hunsaker had exaggerated the opposition to him among aviation operating personnel and urged him to take the job as head of the Material Division, although Land was "willing to do whatever is for the best interests" of the navy and left the final decision to Admiral Taylor. Hunsaker got his way. Taylor convinced Land to take the position, with Hunsaker slotted as head of the Material Division's Design Section.[47]

The arrangement was all Hunsaker could hope for, generally carrying over the organization and personnel he had so carefully cultivated during the war. Garland Fulton continued in the position of executive officer, with dual responsibility for the aircraft design subsection. Starr Truscott headed the lighter-than-air subsection. Lt. Comdr. Sydney M. ("Mose") Kraus, a friend and Annapolis classmate, came from the Bureau of Steam Engineering to handle aircraft and engine procurement. Another newcomer was Bruce Leighton, now a lieutenant commander, who was Hunsaker's choice to head the engine design subsection. Confident that these men could deal with much of the day-to-day routine within the section, Hunsaker said that he was then free "to run the show as an extra number with time to dig into any problem that needs attention."[48]

Before the year was out, Hunsaker received news that he was in line for promotion to commander. On 17 December 1921 he began a two-day series of physical, written, and oral examinations, culminating in notification that the selection board had approved his promotion. The formal date was 31 December. Along with professional advancement, there were changes at home during those years in Washington. One of the most significant came on 24 September 1919, when the Hunsakers bought their first home. Occupying two lots at 3515 Lowell Street in northwest Washington and costing $22,000—a substantial sum in the days when a new luxury car could be had for about $1,000—the house was a handsome two-story brick colonial with a fine prospect across the street to the National Cathedral. Alice loved the garden, and the children had plenty of room to play in the large yard. Hunsaker himself worked hard but still found time to take fishing trips, attend football games in Annapolis, relax with the occasional novel, and even join the new Congressional Country Club, just outside the district in Bethesda.[49]

By the end of 1921, three years filled with turmoil and anxiety ended and a degree of stability came to naval aviation in general and to Hunsaker's life and career in particular. In helping to delineate the administrative and operational structure of naval aviation, he had achieved two principal objectives: ensuring the continuation of aviation as an integral component of the fleet and preserving the role of aviation technical officers in the reorganized bureau structure. He had also enjoyed success in outlining priorities for postwar aircraft design and construction and had ensured that the United States had access to most, if not all, of the latest aerodynamic and structural information from Europe. The loss of ZR-2 had been a tragic setback. Hunsaker thought he knew what had gone wrong and was reasonably confident that such mistakes would not be repeated, but lingering doubts remained about the navy's rigid airship program. Billy Mitchell's continuing opposition to the navy and its aviation arm was worrisome, too. Hunsaker remained confident that the new Bureau of Aeronautics and its technical officers had a far better understanding of aviation's role in the nation's defense than did the air power advocates. Despite uncertainties, Hunsaker could look back with satisfaction that the part he had played in the technical and administrative reshaping of naval aviation in the years after the war provided a solid foundation for the future.

5.   Service at Home and Abroad

For Jerome Hunsaker the multifold challenges of reorganizing naval aviation and orchestrating peacetime aircraft development were only part of the tasks facing him as naval aviation underwent changes in the early and mid-1920s. Still ahead for the navy's air arm was establishing and maintaining its technological and operational competence and, most important, heading off the recurring political threats of those like Billy Mitchell, who wanted to see naval aviation subsumed into an all-encompassing unified air force. All this had to take place within a naval establishment greatly reduced in size and under severe fiscal constraints. Hunsaker's passion for hard work, plus his technical know-how, political savvy, and growing circle of influential friends in and out of the service—all marks of the heterogeneous engineer—helped the navy meet those challenges.

Few projects occupied as much of Hunsaker's time as did the American-designed and -built airship ZR-1, work on which had been held up in early 1921 due to problems in duralumin fabrication and delays in the construction of the Lakehurst hangar. Putting the extra time to good use, Hunsaker and his staff modified the layout of the engines on the airship and worked an additional gas cell into the design, lengthening the hull from 677 feet (206 meters) to 710 feet (216 meters) and increasing its volume to more than 2 million cubic feet (56,634 cubic meters). By the summer of 1921 the design was well enough in hand and materials deliveries had caught up sufficiently for ZR-1 fabrication to begin at the Naval Aircraft Factory. In the meantime, the Lakehurst hangar neared completion, where the ZR-2 was to be berthed and final assembly of the ZR-1 would take place.[1]

Then, in August, came the ZR-2 accident. In the aftermath of the tragedy, Hunsaker was anxious to have as much information as possible on the structural deficiencies of the British airship while it was still possible to make changes in the ZR-1. Charles Burgess recommended abandoning the ZR-1 project altogether and starting over. Hunsaker disagreed. He told Burgess in March 1922 that it was "entirely out of the question" to design and build a new airship and that because of the navy's substantial financial investment it was too late to abandon the project. "The only way that I can see to make a go of airships," Hunsaker argued, "is to go ahead on ZR-1 and make a good job of it." Put another way, "the situation as regards the future of airships entirely hinges on what we do with ZR-1," and if it failed, Congress would not give them a chance to build any more.[2]

To ensure that the ZR-1 project succeeded, Hunsaker needed information from Burgess on British and German construction methods and any insight he had into the reasons for the ZR-2 crash that might be directly applicable to the ZR-1 design. He was most concerned about the low factor of safety built into the ZR-1 design and wanted Burgess, who was still in Britain, to return home and help with a series of reports on the structural integrity and anticipated performance of the airship. The information would be made available to a committee of the National Advisory Committee for Aeronautics charged with reviewing the airship's design and construction. The Investigatory Committee had been created at the request of Admiral Moffett, who agreed with Hunsaker that the navy's airship program hung in the balance and that immediate action was needed in the wake of the ZR-2 disaster.[3]

The Investigatory Committee, chaired by Henry Goldmark, a civil engineer, and including William Hovgaard and Max Munk as members, went about its business thoroughly and professionally, examining all aspects of the airship's design and construction. But it was clear from the beginning that Hunsaker thought little could be gained by "further argument and debate about the structural design of the hull," major changes to which would incur further delay in the ZR-1's construction. Back in the United States, Burgess provided the committee with memoranda on stress calculations, weight distribution, and power estimates. Even though there were no basic alterations, changes were made to the ZR-1 design in the spring and summer of 1922, the most important of which were modifications to the control system to limit stresses in high-speed maneuvers and fitting stronger longitudinal girders in the aft part of the airship. In its final report, issued on 4 December 1922, the Investigatory Committee found no reason to suspect that the ZR-1, unlike its British counterpart, had any structural flaws and concluded that the navy's design was based on the best available data obtained through meticulous research and testing.[4]

According to Hunsaker, the ZR-1 got "a clean bill of health" from the committee, yet the investigation was far from an impartial exercise. Hunsaker himself was a member of the NACA's Main Committee, appointed in October 1922.[5] Nominally independent, the Investigatory Committee was made up of people who knew Hunsaker personally and respected his experience in the field. Munk, in particular, would have found it hard to disagree, knowing that he owed his job at Langley to Hunsaker, and Hovgaard was an old MIT associate. Moreover, Hunsaker had made it clear from the beginning that the committee had a free rein to look into all aspects of the airship project, provided they found nothing that might require a complete redesign or threaten the airship's cancellation. The committee's investigation was Hunsaker's means of reaffirming that his original decisions about the airship's design were correct and that he was still headed in the right direction.

So the ZR-1 project continued as before, even though Hunsaker and others had lingering doubts about the airship's strength. The Naval Aircraft Factory completed fabrication of the main hull components in the early fall of 1922 and finished shipments to Lakehurst by the end of the year. On at least two occasions in late December 1922 and late February 1923, Hunsaker made inspection trips to Lakehurst to observe hull construction techniques and gas cell inflation tests. As the ZR-1 neared completion in the summer of 1923, Admiral Moffett decided to send Hunsaker and Burgess to Friedrichshafen to inspect an airship being built for the navy by the Zeppelin company.[6]

German construction of an airship for the United States Navy had come about through the Treaty of Versailles in 1919, which stipulated that the United States was to receive two German airships as spoils of war. Before the Americans could take possession, however, the airships' crews destroyed their craft. Even though Washington was not a signatory to the Versailles treaty, the former allies determined that the United States deserved compensation and directed the Germans to build a new airship as restitution. Garland Fulton received orders transferring him to Friedrichshafen as the Bureau of Aeronautics inspector of naval aircraft at the Zeppelin works, where he was to work with the Germans in drawing up the general specifications for the airship. Hunsaker believed it would be "an interesting job" and that Fulton's experience in airship design "entitled" him to it, but Fulton's departure at the end of February 1922 meant that Hunsaker had to find a new assistant. His choice was Lt. Comdr. Walter Wynne Webster, a Naval Academy graduate and Construction Corps officer who was a shop superintendent at the New York Navy Yard and enthusiastic about becoming involved in aviation. Hunsaker told him that he wanted him "to come down here and let me teach you to play the piano and then to tune it and make its action smoother."[7]

Fulton, meanwhile, was in his element in Friedrichshafen, leaning over the shoulders of the Luftschiffbau engineers and assiduously noting all relevant details of their design and construction techniques. He forwarded drawings and data on the airship—designated the ZR-3—to Hunsaker, who used them to make additional last-minute improvements to the ZR-1. Hunsaker realized that Fulton had to tread lightly in Friedrichshafen. The Germans jealously guarded their airship secrets but knew they had no choice other than to cooperate with the Americans. For his part, Hunsaker was willing to share enough information about the navy's airship work to convince the Germans that the Americans were proceeding independently and to gain their confidence in hopes of learning more about their materials and design concepts. He also wanted to steer the Germans gently into some of the areas he thought were important. He wrote to Fulton in January 1923 that "I am pretty well satisfied that we have the matter of general structural strength calculations on a firm foundation," and he urged Fulton to see that the Germans compiled aerodynamic data to correlate with those strength calculations before proceeding further with the ZR-3 design.[8]

Hunsaker joined Land and Sydney Kraus aboard the United States Line's *President Harding* on 9 June 1923 bound for Europe and another look at foreign aviation developments. In Friedrichshafen Hunsaker for the first time met Hugo Eckener, the brilliant and charismatic head of Luftschiffbau Zeppelin, whom he remembered in later years as a "wonderful man." Eckener understood what his company had at stake in the ZR-3, for it might well be the last airship the company would build and that to ensure any future for the firm it had to be of exemplary design and construction. In Paris Hunsaker worked with the attaché's office to gather intelligence on the latest French aeronautical activities. Then he flew to London, spending time with his friend Jack Towers and having dinner with Griffith Brewer, a lawyer, patent agent, and friend of Orville Wright. He returned to the United States aboard the steamer *President Roosevelt,* arriving in New York early on the morning of 3 August after an eight-day crossing.[9]

Back in the United States, Hunsaker and others in the Bureau of Aeronautics anxiously awaited the ZR-1's first flights. During the second week of August, Hunsaker spent two days at Lakehurst while the airship's gas cells were inflated and hangar tests were conducted. A few days after the ZR-1's first flight on 4 September 1923, Hunsaker received a congratulatory letter from Ralph Upson, now with the Aircraft Development Corporation in Detroit. Upson said that it was satisfying "to see the ship after the years of work we have spent on its design and construction performing so satisfactorily. . . . I think that results to

The country's first rigid airship, the ZR-1. Later named the *Shenandoah,* it took to the air for the first time in 1923, after many delays in its construction. The airship was another product of Hunsaker's Aircraft Division design team. (87-9680, Courtesy NASM)

date are sufficient to indicate that she may be relied upon to give a good account of herself." On the twentieth, Secretary of the Navy Edwin Denby commended Hunsaker for his "ability and untiring zeal" in the ZR-1 project and extolled the airship as a key addition to the defense of the United States.[10]

The ZR-1 had been built at the Naval Aircraft Factory, the government-owned and -operated facility at the Philadelphia Navy Yard. Not long after the creation of the Bureau of Aeronautics, the navy and Hunsaker faced a growing controversy as the American aircraft industry underwent a severe contraction in the years after the war and pressure increased to limit the factory's aircraft manufacturing activities or to close it down entirely. The Material Division was chiefly responsible for the operation of the Naval Aircraft Factory, with engineers constantly shuttling back and forth between the bureau and Philadelphia. From time to time, Hunsaker himself had been assigned to temporary duty at the factory, working with the factory's design staff on a variety of projects during and after the war. On one hand, he wanted to see the factory's aircraft manufacturing continued, because it provided practical experience for engineering personnel and kept a core of experienced workers in place. In 1920 he told George Westervelt, now a captain and manager of the factory, that "it is pretty hard to see how you can keep your peace of mind while running a great number of designs and experimental machines through with no backbone of production to keep the force steadily employed." On the other hand, he sympathized with private industry, which relied on the government for orders and faced extinction in a rapidly shrinking postwar market.[11]

In any case, manufacturers were not happy with the status quo, considering the Naval Aircraft Factory unacceptable government competition in an industry in free fall. A shrewd politician, Admiral Moffett immediately comprehended the seriousness of the problem. He could not very well shut down the factory without incurring the wrath of the technical officers in the Bureau of Aeronautics, not to mention Pennsylvania's congressional delegation, which might retaliate by voting against future aviation appropriations. But he understood that he had to effect economies in the new bureau while at the same time assuaging the sensibilities of the manufacturers, on whom he would have to rely for aircraft both now and in time of a national emergency.[12]

In the fall of 1921 Moffett asked Hunsaker what he thought would happen if the navy eliminated all aircraft manufacturing at Philadelphia. Hunsaker replied that it would not save any money, at least for the immediate future. Some 300 people would be needed just to store unused equipment at the factory, contracts would have to be signed with private manufacturers to complete unfinished airplanes, and existing stocks of raw materials, parts, and supplies would have to be disposed of at bargain basement prices. Most important, the navy might find itself unable to obtain at reasonable cost the types and numbers of aircraft it needed if it had to rely totally on private manufacturers for all of its requirements. "In general," Hunsaker maintained, "abandoning manufacturing at the Naval Aircraft Factory will result in no economy to the Government." Far more serious than the economic repercussions would be the loss of personnel at the factory: "The organization which has been built up over a period of years will be disbanded, with no hope of ever reassembling them," he told Moffett. If worse came to worst, and the navy had to pare its aviation expenditures to the bone, Hunsaker recommended placing all of its aircraft orders with the factory and letting the private firms wither on the vine.[13]

Moffett chose a compromise. In January 1922 he promised aircraft manufacturers that "it is not the policy of the department to go into production of aircraft at the Naval Aircraft Factory." The plant would continue its wide range of design, research and development, and testing, but in the future it would no longer undertake major aircraft production. Instead the factory would concentrate on manufacturing only a handful of experimental models "in order to keep a check on costs, time of construction, together with a general line on commercial activities." Although it did not fully satisfy the manufacturers, Moffett's compromise proved to be a workable solution to the problem, defining the functions of the factory throughout the remainder of the decade.[14]

Flight training for the bureau's technical and administrative officers was another issue at the time. The legislation creating the Bureau of Aeronautics in 1921 stipulated that the bureau chief and other senior officers had to be qualified as aviators or aviation observers. The naval aviation observer billet came about as a means of heading off a last-ditch effort by Billy Mitchell to prevent Moffett's appointment by adding the proviso that the bureau chief must be an aviator. Mitchell knew that Moffett was not a flier and expected that at this late date he was unlikely to pass a rigorous flight training course. Clever political maneuvering, however, resulted in wording in the legislation permitting the bureau chief to be a "naval aviation observer," whose job was to assist a pilot during long reconnaissance flights and who did not have to undergo as demanding a training regimen as did a full-fledged naval aviator.[15]

On 1 May 1922 Moffett, along with Hunsaker, Land, Kraus, and six other engineering officers, received orders to report to Pensacola, Florida, for observer instruction. All had to undergo a physical examination. Hunsaker remembered his experience well. Dr. Cary T. Grayson, formerly the personal physician to President Woodrow Wilson, conducted the physical, which "included whirling round in a dental chair, after which I lost my breakfast." He survived this unpleasant test for motion sickness but failed the examination because of poor eyesight and substandard hearing due to chronic otitis. In later years he reasoned that he never would have made a good aviator and, thinking of the numbers of men who died in accidents testing new aircraft, concluded that the rejection "possibly saved my life." Those who passed the physical, including Moffett, Land, and Kraus, went on to the five-week course in Pensacola, earning their observer's wings in June 1922.[16]

Among Hunsaker's many responsibilities in the Bureau of Aeronautics Design Section was seeing that the navy fielded competitive entries in the numerous air races held in the early 1920s. Like most engineers at the time, Hunsaker believed that racing "improved the breed," allowing promising new designs to be tested in a real-world "laboratory" while showcasing the navy's latest aircraft and best aviators. Hunsaker's interest in the development of racing aircraft came about in part as a result of the disappointing results of the American effort in the Gordon Bennett races held in France in September 1920 under the sponsorship of American newspaper publisher James Gordon Bennett. The French easily won—their third victory in a row—and retired the Gordon Bennett Cup. Hunsaker concluded that the French victory was the result of an efficient organization and well-tested airplanes, and he was amazed that anyone seriously thought it possible to compete as the Americans had with new and virtually untried aircraft.[17]

If the Gordon Bennett races indicated to Hunsaker that reliability was more important than speed, the Pulitzer Trophy race later in the year provided convincing evidence that proven designs offered more chance for success. Held under the auspices of the Aero Club of America, the races took place at Long

Island's Mitchel Field on Thanksgiving Day, 25 November 1920. Hunsaker attended as a member of the Aero Club's Technical Committee. The army's powerful Verville racer was victorious, but the airplane did not maintain as high a speed as many thought it would, and its engine had to be operated at reduced horsepower to ensure reliability. To Hunsaker, the races underscored the advantages of using mature designs, which had undergone rigorous prerace testing and evaluation. A high percentage of the Liberty-powered army DH-4s finished, as did many of the navy's VE-7s.[18]

Anticipating the 1921 Pulitzer races in Omaha, the navy ordered two CR racers from Curtiss, taking delivery of the airplanes just before the services, by mutual consent, decided not to compete in any races that year. In November Curtiss entered one of the CRs in the Pulitzer race, where it easily won over a small field of private entries. The CR's success and tests later in the year convinced Hunsaker that it was a good idea for the two services to get back into racing. In February 1922, he wrote to Maj. Thurman H. Bane, commanding officer of the army's airplane engineering division at McCook Field, asking for "some dope as to what the Army possibilities are for the Pulitzer Race," and telling him that he was trying to "jazz up interest" in the navy for a return to racing. Competition, Hunsaker told Bane, "is good policy for stimulating the development of planes and engines . . . and the more interest we can get in this thing the better it is for all concerned." Bane replied that the army was "studying the matter" but had not yet made any decisions about reentering competition.[19]

In part because of Hunsaker's insistence, the navy devoted maximum attention in 1922 to the Curtiss Marine Flying Trophy (a seaplane competition) and the Pulitzer races (for land planes). The Detroit Aviation Society sponsored both races, with the Curtiss event held in Detroit proper in early October and the more prestigious Pulitzer coming off a week later at Selfridge Field north of the city. The Curtiss race allowed Hunsaker to compare the performance and reliability of air-cooled radial engines with water-cooled inline engines. In preparation for the competition, the navy fitted one of the new TS-1 shipboard fighters (designated the TS-2) with a water-cooled 180-horsepower Aeromarine engine and another (designated the TR-3) with a 200-horsepower Wright water-cooled engine. Two additional airplanes (designated the TS-1 and the TR-1) retained their 200-horsepower Lawrance radials but had different wings. The Lawrance-powered TR-1 won the race on 8 October, verifying the inherent robustness of the fighter's design and construction and the superiority of the air-cooled radial in naval applications.[20]

For the Pulitzer the navy attacked on two fronts. First, the veteran CRs (redesignated CR-2s) were rebuilt with modifications; one of the airplanes was fitted with new low-drag radiators faired into the wings. Second, Hunsaker was given a blank check to design and build two new "mystery" racers in collaboration with the Wright company, which had recently established an airplane division in Long Island City, New York. Relying on his knowledge of European aerodynamic developments, Hunsaker adopted the sesquiplane configuration, pioneered by the French and featuring a small half-wing suspended beneath the fuselage. The airplanes, designated the NW (for Navy-Wright), were powered by 525-horsepower Wright T-2 water-cooled V-12s and featured Hunsaker-designed rubber shock absorbers in their landing gear. Despite Hunsaker's best efforts, it was impossible to complete both airplanes in time for the race on 14 October. The results were a bitter disappointment for the navy team. Army Curtiss R-6s came in first and second, with the navy CR-2s taking third and fourth. The lone NW had to ditch in Lake St. Clair when its engine seized on the fourth lap.[21]

Hunsaker spent more than two weeks in Detroit and at Selfridge Field providing advice and assistance with the preparation of the aircraft and attending both races with Moffett and other navy brass. The Curtiss race was an all-navy show. Hunsaker was happy with the performance of the TR-1 but was disappointed with the number of racers that failed to finish. "Our friends in the Army," he told a fellow officer in early 1923, "went out of their way to point out that if the Navy went to war they might expect one out of eight of their machines to finish a given trip." Attributing the poor results to the prevalence of untried engines in the race and to problems with floats and propellers, Hunsaker concluded that racing was meant to reveal weaknesses in design and construction and to demonstrate where improvements should be made in high-performance aircraft. He thought it was healthier for the navy to focus its attention on a few airplanes and on races in which there was "real competition even to the extent of our losing."[22]

The "real competition" Hunsaker had in mind was the Schneider Trophy race to be held in England in the summer of 1923. A prestigious international competition for seaplanes, the Schneider promised to be an ideal opportunity for the navy to rise above interservice rivalry and demonstrate its latest technology to the world. Hunsaker broached the idea to Admiral Moffett, who was immediately receptive, promising to provide money and personnel. Redesignated as CR-3s, the CR-2s received floats, new wing radiators, duralumin propellers, revised engine cowls, and larger vertical stabilizers. The NW underwent even more extensive changes, constituting what amounted to a complete rebuild, emerging as the NW-2. No longer a sesquiplane, it was now a conventional biplane with floats, wing radiators, a three-blade propeller, and a new, even more powerful Wright T-2 engine. Unfortunately during trials four days before the

28 September competition, the NW-2's engine disintegrated in flight, sending the craft plummeting into the water at more than 200 miles (322 kilometers) per hour; miraculously, the pilot escaped unhurt. The all-but-obsolete CR-3s saved the day for the navy, finishing first and second in the race, well ahead of the third-place British entry.[23]

Victorious in the Schneider race, the navy shifted its attention to the 1923 Pulitzer in St. Louis. The navy's entries were Wright F2W-1s, powered by 600-horsepower T-3 engines, and two Curtiss R2C-1s, with 500-horsepower D-12s. Hunsaker was in St. Louis for the races, arriving the week before to help with the preparation and testing of the sleek new racers. Postponed by bad weather from the first until 6 October, the Pulitzer was another navy triumph, with the four airplanes taking the top four places. The winner was one of the R2C-1s, which finished at an average speed more than ten miles (sixteen kilometers) per hour faster than the speediest F2W-1.[24]

After the stunning victories of 1923, the navy pulled out of competition, having accomplished all it set out to do. Racing was at best a mixed bag. As a public relations exercise, it brought valuable attention to the service at a critical time of reduced funds and opposition from the army. As a demonstration of advanced technology, racing yielded some improvements in power plants and fuels but precious little in the way of anything that would have a direct, practical application to service aircraft and operations. Nevertheless, competition kept Hunsaker and his staff busy through what might otherwise have been fallow years and helped them keep their edge in the design and development of high-performance aircraft.

Hunsaker also regarded competition as part of a broader scheme of basic research in aeronautics. In October 1922 he was selected a member of the NACA, replacing Admiral Taylor as the navy's technical representative on the Main Committee. Since the NACA's creation in 1915, Hunsaker had coauthored the organization's first technical report, helped define its research agenda in light of European developments, and collaborated with the organization's members on research projects. Matching his background and expertise, Hunsaker served on the NACA's important Aerodynamics Committee and Materials for Aircraft Committee. During the planning stages of the 1923 Pulitzer races, Hunsaker was a member, along with George W. Lewis, the NACA's director of research, of the National Aeronautics Association Contest Committee.[25]

Before leaving for St. Louis and the Pulitzer race, Hunsaker received orders on 14 September detaching him from duties with the Bureau of Aeronautics and sending him to London for service as an assistant naval attaché for air. Collaterally, Hunsaker was to hold down similar posts in the American embassies in Paris, Berlin, the Hague, and Rome. Moffett hated to lose Hunsaker, an offi-

The Hunsaker children in 1922, at home on Lowell Street in Washington: the twins, Jerome Jr. and Peter, the oldest, Sarah, and the youngest, Alice (dressed in a christening gown). (Courtesy Alice Hunsaker Bird)

cer "of exceptionally sound judgment and ability" whose "services have been invaluable to the Bureau." Hunsaker himself was not eager to leave Washington. He was happy in the big Lowell Street house, and the family had expanded with the birth of another daughter, Alice Maclaurin, on 10 August 1922, her middle name given in recognition of the close friendship the family had long enjoyed with the Maclaurins of MIT. Burdened with the responsibilities of wife and mother, Alice preferred to stay in Washington, but she recognized that in a sense she was also married to the navy and that the London assignment was part of her duty, too. Despite reservations about leaving a job he had almost single-handedly created, Hunsaker looked at the new posting as an ideal way to keep close watch on the progress of the ZR-3 in Germany, survey the European aeronautical landscape, and restore some of the feel for basic aerodynamic research that he had lost over the past six years.[26]

In anticipation of leaving the Bureau of Aeronautics, Hunsaker wrote a comprehensive narrative of the development of naval aviation, complete with documents and appendixes. The topics included the historical background of aeronautics, European experiments, wartime design and procurement, administration, the Naval Aircraft Factory, duralumin, helium, and rigid airships.

Fully conscious that the navy lacked much of an institutional memory, Hunsaker saw the history as a means of setting the record straight on matters pertaining to the navy's work in aviation, especially during the war years; it has been of inestimable value to historians ever since.[27]

Before handing his bureau job over to Comdr. Holden C. Richardson, Hunsaker also provided the director of naval intelligence with a preliminary appraisal of the aeronautical situation overseas. Only in France and Britain were substantial sums of money being spent on aeronautics, with the French doing well putting into service standardized aircraft designs and the British leading in air-cooled engines and aircraft design theory. The most egregious French weakness was the inadequacy of its aircraft engines, which Hunsaker attributed to the failure of the government to provide engine test facilities and to a "ring" of manufacturers who had limited competition in the industry. The aircraft carrier was "the outstanding naval development" by the British, although they, like the Americans, struggled with ways of getting the maximum number of airplanes off the carrier's deck in the minimum amount of time. Hunsaker believed the feud between the Royal Air Force and the Royal Navy over control of naval aviation had "impeded development of naval types of aircraft during the past two years," but he thought a compromise between the two services would likely resolve the situation. "Effectively stifled" by the Treaty of Versailles, Germany had excelled in the development of airships—"a brilliant exception" to an otherwise dismal situation. Japan Hunsaker dismissed as little more than a "customer" of France and Britain.[28]

There was other unfinished business before leaving Washington. Hunsaker went to Lakehurst on 10 October for the commissioning ceremonies of the ZR-1, which was named the *Shenandoah*. Meanwhile, a family friend in England, Maj. Philip L. Teed, assured him that the British schools would be fine for the children and offered to have an agent look for a furnished house. Hunsaker later thanked Teed for his help and for having written to Comdr. John Towers "to get quarters for us all at some hotel till we can look around. I think we had better take a house in town until next Spring or Summer and, in the meantime, have a chance to look around at leisure." Before sailing for England, Hunsaker attended the annual Construction Corps dinner in New York on the evening of 23 November, where he said his goodbyes to many of his fellow officers. The next day he met Alice and the youngsters, who had come up on the overnight train from Washington, and they boarded the liner *President Roosevelt* at noon, bound for Plymouth, where they arrived nine days later, on the morning of 2 December.[29]

Attaché duty was perfect for both Hunsaker and the navy. At the time, American naval officers in London and other world capitals had a hard time meeting their professional and social obligations from the "ridiculously small" allowances they received from the navy. Already well on the way to financial independence as a result of Alice's income and his own salary and investments, Hunsaker did not have any worries about covering the additional expenses he knew came with the London posting. Furthermore, he already had worked with Jack Towers and other officers in the London embassy, had extensive personal and professional contacts in the European aeronautical scene, and from experience knew how best to extract information from those sources.

The work of an attaché was strictly quid pro quo; that is, to obtain information it was necessary to provide information. The Office of Naval Intelligence frowned on sleuthing and stressed that attachés use "reputable business methods" in their work, which was just as well, because Hunsaker did not see his job as spying. His first boss in London, Capt. Charles L. Hussey, reported that "he understands the character of the information needed; knows how to obtain it; and is able to prepare concise reports setting forth the information obtained. He naturally makes useful friends. In every way he is a most valuable officer for this sort of duty."[30] Above all, Hunsaker was convinced that the unfettered exchange of scientific and technical information was essential to progress in aeronautics, and he viewed his new job as a means of restoring the dialogue that had been interrupted by the war.

Towers, on his second tour as an air attaché in London, arranged for the Hunsakers to let an apartment at Kensington Palace Mansions. Hunsaker wrote to his father in December: "It is not very good, but is just opposite the Kensington Gardens, where the children can go every morning and again every afternoon. I think I wrote of their fun in discovering the Round Pond, the Serpentine, and Peter Pan's island and even his statue." Despite its shortcomings, the place at least had the advantage of being close to the American embassy in Grosvenor Gardens. Through Teed, the Hunsakers soon found a spacious house on Cheyne Walk in Chelsea, which they rented in early January 1924 from a Royal Air Force officer who had been sent out to Egypt. Alice deemed the house "delightful, and well arranged for a family." Among its most attractive amenities were a drawing room with a "beautiful outlook across a strip of park to the river" and a cozy study where Hunsaker could retreat when he needed to. The Hunsakers also rented a fifteenth-century country house not far from Salisbury, in Wiltshire. Known as Porch House, it was in the Tudor half-timbered style and had a great hall with a huge stained glass window taken from Salisbury Cathedral during one of its renovations. According to Alice, the family "loved the old house," which provided welcome deliverance from the hurly-burly of London and allowed them to enjoy the "delicious" English countryside.[31]

Then as now, a large part of the attaché job involved entertainment. Of primary importance to Alice was finding a good cook and servants to maintain the house and help her meet the family's social responsibilities. To her the London social life was "rather a blur." Every year, along with the rest of the diplomatic corps, they attended the royal court at St. James's Palace, where, seated on thrones, King George V and Queen Mary oversaw a "sea of elegant people" dressed in formal attire. Part of the ceremony was the presentation to the royal family. In turn each lady negotiated a long red carpet, listened for her name to be called by the lord chamberlain, curtseyed before the queen, and made a carefully practiced turn before retreating on her husband's arm.[32]

With her husband occupied at the embassy and often out of the country on official business, Alice centered her attention on the household and family. "Little" Alice, only a year and a half old, needed a nurse, and the other children had to be settled in schools. For the first couple of years, Jerry Jr. and Peter attended the Gibbs School on Sloane Square, within walking distance of home; Sarah (always "Sally" to the family) went to Birtwhistle, a girls' school next door to Gibbs, before transferring to the Graham Street School, which the Hunsakers thought was more academically oriented. The twins spent their last year at the Durnford boarding school in Dorset. Any reservations Alice or her husband had about sending the twins away from home were mitigated by the then-common belief in England that boarding schools were an antidote to the prevailing culture of "child worship" and deterred parents from spoiling their offspring.[33]

Among Hunsaker's routine duties was compiling statistical data on European air services, making biannual reports on aviation personnel numbers, estimating reserve forces, and making sure that articles, technical reports on foreign aeronautical developments, and parliamentary debates on aviation reached the right people in the Office of Naval Intelligence and the Bureau of Aeronautics. Personnel statistics were important indicators of trends and provided a good means of comparison among the various services and with the United States. Understanding the personal characteristics of individuals was essential to painting the intelligence picture, so Hunsaker's and other's impressions were important in making up profiles of key people in aeronautics. Hunsaker also maintained a liaison with the military attachés at the embassies in the countries for which he was responsible. A measure of the lack of trust the services had for each other was that the army's Military Intelligence Division kept a file on Hunsaker, along with a number of his reports, regarding him as a "suspect."[34]

Travel included visits to aeronautical and naval installations and meetings with Hunsaker's military and civilian counterparts throughout Europe. During the first few months of 1924, he confined his trips to England and Scotland, visiting the National Physical Laboratory in Teddington, the Royal Dockyard at

Chatham, and shipbuilding facilities in Newcastle and Glasgow. In June he spent the better part of a week looking at aviation manufacturing plants and research operations in France. The following year saw more trips to Paris and in July 1925 a visit to the Aerodynamics Institute in Aachen.[35]

Even though he had resigned from his position on the NACA's Main Committee, Hunsaker kept up his ties with the organization while he was in London. George Lewis relied on Hunsaker to supplement information from the committee's Paris office, headed by John Jay Ide. Ide's office had become a source of friction between the NACA and military and naval attachés, who thought it overlapped their responsibilities. Hunsaker knew Ide from the Aircraft Division during the war, and he assured Lewis in February 1924 that his and Ide's activities complemented one another. He explained to Lewis that "Ide's freedom of movement in England is pretty closely restricted" and that he had had to secure permission from the Air Ministry to visit aviation manufacturing plants. Some attachés worried that the ministry might eventually limit access for everyone, so Hunsaker made sure that Ide went through the attachés at all times so as not to "draw too heavily on their capital." He foresaw no trouble getting to see what the Americans wanted, although British regulations seemed to spread a "thick blanket over all scientific and research personnel" and restricted the free flow of aeronautical information. Another example of how Hunsaker helped came in 1925, when he intervened to assist the NACA in locating new office space in Paris, thereby allowing Ide to stay close to the embassy and its attachés.[36]

With the ZR-3 nearing completion in the late summer of 1924, Hunsaker was off to Friedrichshafen to observe and report on the ground tests and early flights of the new airship. He arrived at Luftschiffbau Zeppelin on 3 September, too late to see the ZR-3's maiden flight on 27 August, and he was unable to go along on the airship's second flight on 6 September. Like others who had seen or flown in the new airship, Hunsaker was favorably impressed with the quality of the Germans' work and their thoroughness in preparing the ZR-3 for delivery. He left Friedrichshafen on 11 September, the day the ZR-3 made its third flight, and went on to Zurich and Paris, spending a week catching up on the latest French aeronautical work before returning to London on 18 September. A month later, the ZR-3 flew to Lakehurst—the fifth aerial crossing of the Atlantic—and was subsequently commissioned into the navy as the *Los Angeles*.[37]

Not long after he got back to London, Hunsaker drafted a series of reports on European naval aviation to be used by American officers preparing to testify before the Lampert committee on problems in American aviation. Congressman Florian Lampert, a Progressive Republican from Wisconsin, chaired the committee, which first met in March 1924 to probe allegations of an "aircraft trust" dominating military aviation procurement. By fall the investigation had gone far beyond its original purpose and had centered on Billy Mitchell's charges that the United States had fallen behind the rest of the world in aeronautics and needed a unified and independent air force to catch up.[38]

Capt. Luke McNamee, who had recently succeeded Hussey as the naval attaché in London, believed that the British example formed the core of the argument against a unified air force and, using information from Hunsaker, compiled a report for the Office of Naval Intelligence in October. Since naval aviation had been brought under the centralized control of the Royal Air Force in 1918 there had been almost continuous controversy. Arguments that a single air service would save money had not been borne out by the British experience. Most important, in the United States the army and navy had distinctly different tactical and strategic requirements in aviation, which could be met only by separate air arms. "In spite of rapid development of aviation, the Navy is still the Nation's main defense and every effort should be made to keep the Navy at its greatest possible efficiency. This cannot be done as long as the Navy is deprived of control of its aviation," McNamee's report concluded.[39]

A shock to everyone in naval aviation was the loss of the *Shenandoah* in a violent squall over Ohio on 3 September 1925, killing fourteen crew members, among them the airship's commanding officer, Zachary Lansdowne. Immediately, Billy Mitchell went into a tirade, accusing the navy and war departments of "criminal negligence" in causing the accident, and provoking Moffett into angry public outbursts about Mitchell's "unsound mind." Within days, the army brought court martial charges against Mitchell. In the face of the mounting clamor, President Calvin Coolidge created a high-level committee to study and report on all aspects of civil and military aviation. As chairman of the board, Coolidge picked Dwight W. Morrow, an old college chum and internationally recognized businessman. The stakes were so high that the navy ordered McNamee and Hunsaker to return to Washington to testify.[40]

Anxious to know what had happened to "his" airship, Hunsaker asked Charles Burgess for all the "inside dope" on the navy's investigation into the disaster. Burgess said the cause of the accident was a powerful "twister," which subjected the airship to rapid vertical motions and aerodynamic loads that exceeded its structural limits. To Burgess the "horrible thing" was the breaking away of the control car, which caused many of the deaths. The car's immensely strong suspension system had not failed; instead, elevator and rudder control cables had cut through the keel above the car, which collapsed and gave way. Much to Hunsaker's relief, Burgess consoled him that when he had designed the *Shenandoah* no one could have anticipated such severe weather conditions and that even the best meteorology in the world could not have provided enough warning.[41]

The wreck of the navy airship *Shenandoah*. The airship went down in a sudden squall over southeastern Ohio in September 1925, precipitating a bitter fight between the navy and air power zealots led by William ("Billy") Mitchell. (85-19404, Courtesy NASM)

Reassured by experts that the *Shenandoah* tragedy could not have been averted, Hunsaker left London armed with mountains of comparative data to defend the navy before the Morrow Board. He arrived in Washington on 5 October and immediately went to work refuting Mitchell's wild claims that the United States had fallen hopelessly behind other nations in aviation and was vulnerable to attack from a third-rate power. In a long memo to Morrow, Hunsaker specifically countered all of Mitchell's allegations. Japan, which Mitchell charged had greatly increased its aircraft manufacturing capacity, had "no large aircraft production whatever" and was building only about 200 experimental models annually. France, a country Hunsaker knew firsthand, had nothing like the thirty-two aircraft factories Mitchell claimed; rather, the French had only eight "principal firms" and a dozen smaller companies, producing a maximum of 2,500 aircraft a year, far fewer than the 7,000 asserted by Mitchell. As for British aircraft production, it was probably no more than 600 per year, and Mitchell's estimate of 15,000 workers was at least twice the number actually employed. Mitchell's data on Italy's aircraft production were also inflated; the actual figures were about half what Mitchell had stated publicly.[42]

Hunsaker appeared before the Morrow Board on 16 October, the final day of the hearings. He stressed that published statistics on relative air strengths were misleading, usually maximized or minimized for political purposes. In the United States, for instance, the army had greater numbers of smaller airplanes than did the navy, which had emphasized large flying boats. Estimates of aircraft quality were at best an educated guess. Although airplane speed was an essential characteristic, it had to be balanced against other factors, most notably operating ceiling, cockpit visibility, and landing speed. Nor did Hunsaker find any formal mathematical system of weighting numbers of personnel, aircraft, or quality to be workable. In the end, only an observer on the scene who was familiar with the people, the airplanes, and the production systems could render any meaningful judgments about relative air strengths.[43]

From what Hunsaker could ascertain, the United States was "about in the middle" of the leading air powers. The country generally had airplanes that outperformed those of other nations, but it lacked numbers in comparison to the Europeans. In water-cooled engines, seaplane floats, metal propellers, bombsights, catapults, and such mundane but vital aircraft components as engine starters, the United States was superior to any of its potential rivals. In air-cooled engines, superchargers, and metal construction, the Europeans were better. He concluded that the United States had virtually no capability of attacking a foreign nation by air; conversely, an "unsupported" air assault on the United States "seems to be out of the question." Morrow asked about the capability of the long-range heavy bomber, which Mitchell had touted as virtually an irresistible weapon. Hunsaker responded that the maximum range for any airplane was about 3,000 miles (4,828 kilometers), carrying nothing but fuel and the pilot. If the airplane hauled a single 2,000-pound (908-kilogram) bomb, then the maximum practical range fell to 600 miles (966 kilometers). Most telling in light of Mitchell's claims, Hunsaker attested that "no such machine exists today" that could accomplish such a mission.[44]

Back in London by the end of November, Hunsaker received a copy of the Morrow Board report not long after it was made public on 2 December. He was pleased. Even though the report found problems in procurement and personnel, it had "nothing but praise" for how United States naval aviation had been administered, considering it "at least the equal of and in certain directions undoubtedly superior" to that of any other nation. There was no reason that much had to be changed—certainly no reason to combine the navy's air arm with that of the army. In sum, the Morrow Board report was a rebuke to Mitchell and a commitment to the status quo in naval aviation. Hunsaker wrote to Morrow, complimenting him on a "very sound job" that would provide a blueprint for the army and navy "for some years to come." He was also delighted that the *New York Herald,* which had been favorable to Mitchell, "swapped over to the other side in one night" after the Morrow report came out.[45]

In his position in London Hunsaker got along well with Captain Hussey and even better with his successor, Luke McNamee. Both praised his intelligence, initiative, judgment, and reliability. Knowing how valuable he was to their mission, Hussey and McNamee were more than willing to put up with what evidently were Hunsaker's occasional lapses in military bearing, but Capt. William C. Watts, who took over as naval attaché in 1926, was less tolerant. In his fitness reports Watts consistently rated Hunsaker superior in education but only average in aptitude for the service, cooperation, courtesy, leadership, military manner, neatness, physical energy, and subordination.[46]

Watts's reports helped Hunsaker crystallize his thoughts about his future in the navy. Much of his current work did not involve aviation or allow him to exercise his theoretical understanding of aerodynamics. With the *Shenandoah* relegated to a sad corner of history, the new German airship safely delivered, and Billy Mitchell no longer a threat following his court martial and resignation from the army, Hunsaker began to think about alternatives. Before the end of 1926, he would have completed twenty-two years of active service. As a thirty-nine-year-old commander with a superlative record in the Construction Corps, he could look forward to promotion to captain, but beyond that there was only one slot for admiral—as chief of the Bureau of Construction and Repair. That was unlikely. In later years Hunsaker recalled that by the end of 1925 "the Navy looked pretty dull," and he was ready for a change. As for finances, the Hunsakers were relatively well off, with Alice receiving a steady income from her shares in Miss Porter's School in Farmington and Hunsaker himself managing a solid investment portfolio while earning a generous navy salary. But the mid-1920s were boom times, and the private sector beckoned. When Frank B. Jewett of Bell Telephone Laboratories approached him with a job offer and the promise of a free hand in pursuing a new and promising avenue of aeronautical research as a vice president of Bell Labs in New York, Hunsaker, with some reluctance, decided to submit his letter of resignation from the navy on 30 June 1926.[47]

Whatever he had previously thought of Hunsaker, Captain Watts praised his "high professional standing, remarkable fund of knowledge on a great variety of subjects, excellent judgment and zeal in his work," which had "combined to make his services of the very highest value . . . his detachment is deeply regretted." Accolades flowed in from others, too. In accepting Hunsaker's resignation, Acting Secretary of the Navy T. Douglas Robinson wrote,

> The Department regrets that you find it necessary to resign from the
> Navy and will feel the loss of your services. It is hoped that you may
> enjoy prosperity, health, and happiness in civil life and that you will con-
> tinue to hold as close interest and connection with the Navy as may be
> compatible with your civil pursuits.[48]

Hunsaker's separation from the navy was an amicable divorce by mutual consent. He had enjoyed the service, where he had received graduate education in engineering at what became one of the most prestigious schools in the country, had conducted pioneering research in aerodynamics, had met and worked with some of the leading figures in aviation, and had gained administrative experience far beyond what he could have expected in any other job. In return, the navy had benefited from one of the brightest minds in aeronautics, without whom the integration of aviation technology into the fleet would have been far more problematic. Moreover, the divorce was never made final, because Hunsaker retained his links to the navy as a reserve officer and in the years to come answered the call when his old service needed his advice and experience.

6.   Bell Labs

By February 1926 Jerome Hunsaker had made his decision to leave the navy and join Bell Labs. That month he met in London with businessman Harry F. Guggenheim and retired Rear Adm. Hutchinson I. Cone, who were in Britain to take part in a comprehensive study of European aviation for the recently created Daniel Guggenheim Fund for the Promotion of Aeronautics. Harry Guggenheim had joined the legendary First Yale Unit, served with the navy's Northern Bombing Group in France, and been involved in a wartime deal to acquire big Caproni bombers for the navy. Cone, who knew Hunsaker well, had been commander of American naval aviation forces in Europe during the war. Guggenheim and Cone persuaded Hunsaker to assist in their survey by traveling to Italy, where he could kill two birds with one stone: analyze Italian aviation for the Guggenheim Fund and acquire data for the navy on what Italy was doing in the field. Hunsaker's boss, Capt. Luke McNamee, the American naval attaché, sent him to Italy on 25 February for a month-long tour. Several other trips followed, including a five-day visit to Berlin in the middle of July, Hunsaker's last official assignment before his detachment from the American embassy and reassignment to the Office of Naval Intelligence in Washington pending his resignation from the service.[1]

Not coincidentally, the meeting with Guggenheim and Cone came at a crossroads both in Hunsaker's career and in American aviation. By 1926 the postwar era of aviation was over. Aircraft and power plants had reached a plateau of reliability and performance, with few technological breakthroughs likely in the next decade or so. Authorization of the navy's five-year aircraft program in

June 1926 gave Hunsaker every reason to expect that the focus would be on procurement rather than on airplane design or aerodynamic research. At the same time, commercial aviation stood on the edge of rapid, sustained growth, stimulated by the privatization of the airmail under the 1925 Kelly Act and the prospect of new ground and air communications. Hunsaker wrote at the time that "the airplane and engine have reached a commercial degree of reliability, but the technique of operation needs development." Like so much in aeronautics, the airways communications system was invisible, taken for granted by the tens of millions who relied on air transportation to carry them safely and swiftly to their destinations but only just beginning to come into its own. To Hunsaker, airways communication was "the new exciting thing in the mid-twenties."[2] He might have added that airways communication allowed him to transfer what he knew about engineering and the structure of naval aviation to the rationalization and organization of a great new commercial air network.

Hunsaker and his family sailed from Southampton on 19 August, arriving in New York on the twenty-seventh and taking up temporary residence in a hotel. By the beginning of November, they had settled into a rented brownstone at 113 Willow Street in fashionable Brooklyn Heights, a fairly easy subway commute to Bell Labs, which at the time was in a multistory building on West Street in lower Manhattan. Bell Labs was the result of a commitment to industrial research by AT&T, the proverbial "phone company." The company retained a strong interest in aviation not only as heir to the work of Alexander Graham Bell and the Aerial Experiment Association but also because of its extensive wartime involvement in aircraft radio telephony. As president of the laboratories, Frank Jewett, a University of Chicago Ph.D. in physics, was primarily responsible for recruiting promising scientists and engineers to fulfill that mission. He saw an opportunity for AT&T to expand its interests in aviation by developing a communication system whose services could be sold to airmail carriers when they moved, as expected, into passenger-carrying operations. Hunsaker's engineering background, aviation experience, and personal contacts in the aeronautical community made him ideal for the job.[3]

Hunsaker's and AT&T's interest in airline and airway communications meshed with that of Harry Guggenheim and the board of directors of the Guggenheim Fund, who believed that a ground and air network could greatly enhance the safety and efficiency of passenger operations. Guggenheim and Cone began investigating the possibility of setting up a model airline to demonstrate such a system and identified Western Air Express as a prime candidate. Western Air Express was the result of the work of Harris M. Hanshue, a Los Angeles automobile dealer who, with several other businessmen in the city, established the company in July 1925 to bid on one of the new contract air mail routes opened up under the Kelly Act. In the spring of 1926, Western Air Express began carrying mail and passengers over a 700-mile (1,126-kilometer) route between Los Angeles and Salt Lake City via Las Vegas. After less than a year of flying, the company reported a small profit—no mean feat at the time—and Hanshue began thinking about ways of expanding. The Los Angeles–San Francisco route looked attractive, although Hanshue knew that it would be next to impossible to make it a paying proposition without an air-mail contract.[4]

At Cone's suggestion, Hunsaker in May 1927 wrote to Corliss C. Moseley, the operations manager for Western Air Express, for details on the airline's communications requirements. Moseley recognized that communication "must be absolutely reliable and available at all times" for the safe and efficient operation of an airline. Ground stations carried information on weather, dispatching, fuel use, passenger volume, and aircraft maintenance. Telephone and telegraph links, Moseley thought, were superior to radio connections between ground stations but believed that it was "practically prohibitive" for the airline to lease a private wire for such purposes.[5]

Hunsaker used information from Moseley, Western Air Express, and other airlines, as well as his connections within the Guggenheim Fund, to prepare a report on airline and airway communications. In his survey, completed on 1 July 1927, only weeks after Lindbergh's 20–21 May New York-to-Paris flight sparked a nationwide aviation hysteria, Hunsaker determined that the airlines, as they expanded their route systems and began carrying passengers, would need reliable and efficient ground and air communication. He predicted that by 1933 air transport would need at least $500,000 a year for wire services and $2 million a year for radio services. "It seems clear that we are at the beginning of a development which may soon attain substantial proportions," he concluded. AT&T needed to investigate radio equipment as an aid to airway navigation and for the transmission of weather data to aircraft. For a two-way air-to-ground radiotelephone communication system the company should secure suitable radio frequencies, design compact and lightweight transmitters and receivers, and offer the service at a reasonable cost. In the end Hunsaker believed the Bell System would find such services highly profitable.[6]

Following up on his 1 July report, Hunsaker identified Pan American Airways as an airline that might be interested in AT&T's proposed services. Organized by the group of wealthy investors who backed Colonial Air Transport, Pan Am obtained a contract to carry mail between Key West and Havana, with additional assurances of continuing the line to Panama. The mail contract promised to be lucrative, but the route also had considerable appeal to thirsty passengers escaping American Prohibition in Havana's gambling casinos and nightclubs. Pan Am planned to use trimotored equipment, thereby seeming to guarantee that no aircraft would be forced down by engine failure. For communication, the airline intended to rely on two-way radiotelephone connections between Key West and the airplanes aloft and on telephone and cable services between Key West and Havana. Late in the summer Hunsaker met with the charismatic general manager of the airline, Juan T. Trippe, and came away from the discussions convinced that AT&T should take advantage of a rare opportunity to expand its interests into the West Indies and Central America in support of Pan Am's air routes.[7]

Harry Guggenheim considered improvements in weather reporting essential for safe and efficient airline operations. The Weather Bureau's system, intended originally to benefit agriculture, was seriously deficient and did not meet the needs of an expanding air transportation network, whose operations depended on timely reports and precise short-range local forecasts. Consequently, the Guggenheim Fund created the Committee on Aeronautical Meteorology in August 1927 to study the problem. At the suggestion of Lt. Frederick W. Reichelderfer, the chief meteorologist in the Bureau of Aeronautics, Guggenheim chose Carl-Gustaf Rossby, a young and talented Swedish-born meteorologist, to chair the committee. Rossby, an adherent of the revolutionary Norwegian theory of air-mass analysis in weather forecasting, had worked briefly for the Weather Bureau before he found the agency completely unreceptive to what many then thought were heretical ideas and methodology.[8]

On 15 September, Hunsaker met with the committee to discuss ways of obtaining and disseminating weather information along the airways. Meteorologists assigned to each airport on the airway would use the commercial telephone network already in place for ground communications and for gathering and assimilating weather information. Weather reporting stations were to be spaced every fifty miles (80.5 kilometers) along the airway and at locations as far as a hundred miles (161 kilometers) on each side of the route, submitting reports every two hours. Radiotelephones would provide information to aircraft in flight. Rossby suggested that the Guggenheim Fund pay for setting up the weather system on a section of airway as a demonstration project, with AT&T making a substantial contribution to the communications facilities.[9]

Hunsaker liked the idea, comparing the committee's plan to a weather reporting system used by the army in France during the war. The setup had worked well, using the Signal Corps telephone network to relay weather information to a central collection point, from which forecasts went out to units in the field. But he also detected flaws in the committee's scheme. Summarizing the plan for his bosses at AT&T, Hunsaker concluded that it was too "elaborate" and "probably impractical on account of the expense and the personnel required." Regardless of the scale of the project, there was no doubt in Hunsaker's mind that it was beyond the capabilities of the Weather Bureau. "It is, however, desired to organize one model section of airway with complete meteorological service with a suitable communication system and to determine by actual working to what degree such an organization is worthwhile."[10]

Rossby's recommendation received a favorable response from the Guggenheim Fund, which agreed in October to pay for a trial weather reporting system on the New York–Chicago airway, which was more heavily used than other routes and plagued with some of the worst weather of any segment of the national airway network. By the fall of 1928, the airway had radiotelegraph stations operated by the Department of Commerce at Hadley Field, New Jersey (the New York terminus), Bellefonte, Pennsylvania, Cleveland and Bryan, Ohio, and Maywood, Illinois (the Chicago terminus), with operators on duty twenty-four hours a day to provide point-to-point communications. Complementing the radiotelegraph installations were three meteorological stations operated by the Weather Bureau at New York, Cleveland, and Chicago. The usual procedure was for the operators at Bellefonte and Bryan to radio intermediate fields along the route to obtain weather information. Meanwhile, the people at New York, Cleveland, and Chicago used commercial telephone lines to gather additional meteorological data and to correlate it with the information coming in from Bellefonte and Bryan. That information, combined with the normal twice-daily national maps plotted by the Weather Bureau, formed the basis for the forecasts used by the airlines. The system had been pieced together, and it often failed to provide the most up-to-date information.[11]

Rossby's committee asked Hunsaker and Bell Labs to prepare a detailed proposal for a more efficient airway communication and weather reporting system. Recognizing that the system could use equipment already in place and guessing that it would require a minimum of additional research and development, Hunsaker joined with the operating people at AT&T to prepare the plan, which took shape over the next month. The Bell Labs–AT&T scheme relied on an integrated network of radiotelephone stations and long-distance lines to collect and dispatch information from airport to airport and to aircraft en route. Information would be relayed to New York, Cleveland, and Chicago, where meteorologists would analyze it and issue updated forecasts to stations along the route every two hours, compared with the four-hour forecasts obtained previously. Continuous two-way radiotelephone communications would relay weather information to airplanes in the air, which would also use a radio beacon system for navigation. AT&T would lease the complete package for an agreed-upon fee, allowing savings over the usual commercial rates for long-distance telephone and other services.[12]

In early November, Hunsaker met with Harry Guggenheim to make sure that he fully understood what the Guggenheim Fund wanted from AT&T. He learned that the fund was prepared to pay for the project, but beforehand Guggenheim needed assurances that the plan could be used on all airways, not just on the New York–Chicago route, and that if the experiment succeeded the Department of Commerce would push Congress for money to implement a similar system nationwide. Guggenheim also expected AT&T to donate the use of its long-distance telephone lines for the duration of the trial period. At first reluctant to offer its services for free, AT&T agreed to the proposal.[13]

On 13–14 November, Hunsaker and his colleagues met in Washington with representatives of the Department of Commerce and the Weather Bureau to review the general outlines of the Bell Labs–AT&T plan. They agreed to a six-month trial. The government people did point out that mail and express over the New York–Chicago airway totaled $3,000 daily and wondered whether it made economic sense to install a weather and communication system costing up to $1,300 a day to operate. Hunsaker argued persuasively that it was necessary to make an investment in the safety and reliability of the airway to encourage its more intensive use; if the weather reporting system worked well, costs would go down. It might even be possible to encourage local communities to pay for part of it.[14]

Hunsaker's efforts to put together a comprehensive airway telecommunication system also involved the design and development of new ground and air radio equipment, which AT&T hoped to make part of its service to the government and the airlines. The Aeronautics Branch of the Department of Commerce, created by the Air Commerce Act of 1926 and headed by Assistant Secretary William P. MacCracken Jr., made improved radio communications one of its top priorities. In cooperation with the Aeronautics Branch, the Bureau of Standards set up an experimental ground-to-air radiotelephone apparatus at College Park, Maryland, outside Washington, in late 1926, followed by another unit at Bellefonte on the New York–Chicago airway in the spring of 1927. Concurrently, the bureau began developing an improved radio compass that employed new technology to modulate and align the signals sent out by the ground transmitter.[15]

Largely at Hunsaker's insistence, Bell Labs implemented a major radio research and development program. Immersing himself in radiotelephone and radio navigation technology, he studied German World War I radio direction-finding systems, visited College Park, and corresponded with Bureau of Standards researchers involved with government radio projects. By the end of 1927 Hunsaker had recommended that AT&T begin an experimental program using Bell Labs' large and well-equipped radio research facility at Whippany, New

Jersey. Hunsaker did not get far into his work before it dawned on him that it might be possible to use a single transmitter to send weather information and directional signals to a compact radio receiver and "compass" in the airplane. Because of the limited bandwidth available for airway radio communications, the use of one transmitter and frequency for both weather information and navigation would help to minimize interference and improve reception. He also urged the development of an advanced radiotelephone system allowing two-way ground-to-air communication. Following up on these ideas, Bell Labs moved quickly in the winter and spring of 1928 to design and test experimental apparatus, using a leased Fairchild FC-2 single-engine cabin monoplane to carry equipment for measuring the strength of signals transmitted from the Whippany station.[16]

Development of the new radio technology proved simpler than getting everyone to go along with a trial of the AT&T–Bell Labs communications system on the New York–Chicago airway. After securing what appeared to be an understanding with the Department of Commerce and the Weather Bureau, Hunsaker saw the agreement start to unravel toward the end of 1927. The first indication of trouble came from the Weather Bureau. Concerned about costs, the bureau suggested that the system be reduced to cover only the Cleveland-Chicago leg of the route (where there were already many stations serving farmers) and that the frequency of telephone reports be cut from one every two hours to one every four hours.[17]

Harry Guggenheim also had second thoughts about the AT&T–Bell Labs proposal. After discussing the details of a West Coast airway project with Harris Hanshue of Western Air Express in the fall of 1927, Guggenheim was convinced that Western Air Express was well managed and deserved an opportunity to demonstrate passenger operations. Moreover, Guggenheim indicated that AT&T might need to do more on the West Coast as a quid pro quo for the Guggenhiem Fund's support of the New York–Chicago route. He urged Hunsaker to get his company to contribute additional personnel and resources to the Western Air Express weather reporting service and to upgrade the airline's communication system. In an attempt to find a way around the growing impasse, Hunsaker agreed that AT&T could assign more of its people to stations along the Los Angeles–San Francisco route and suggested that it might be more economical to use a teletype instead of the telephone for the two-hour reports.[18]

Not much had changed by January 1928, when Hunsaker wrote to Wesley L. Smith, the president of National Air Transport, which held the New York–Chicago airmail contract, telling him that "I am afraid progress has been backward rather than forward" on the AT&T–Bell Labs scheme. The Weather Bureau, lukewarm about the plan from the start and worried that it would divert scarce resources from its own system, had begun a program to upgrade its service to the airlines with four reports per day using existing communications equipment. As long as the Weather Bureau opposed the idea, it was unrealistic to expect the Department of Commerce to do much on its own. It appeared, too, that the Guggenheim Fund was no longer willing to support the system for the New York–Chicago route; it had sent Rossby to California to look at the operations of Western Air Express between Los Angeles and San Francisco and had lent the company $180,000 to acquire three new twelve-seat Fokker F-10 trimotors. Hanshue and others thought the big, luxurious airliners would attract fare-paying passengers, while Guggenheim saw the deal as an example of the financing that airlines would need as they moved ahead with multiengine equipment.[19]

Neither Hunsaker nor others at AT&T or Bell Labs were particularly keen on shifting their focus to the West Coast. The New York–Chicago airway had more varied weather conditions than did California, and the route was practically in AT&T's backyard, simplifying the logistics of getting the system started and monitoring its operation. Nevertheless, AT&T could not afford to be excluded from a West Coast airway plan, should that prove the way Guggenheim wanted to go. To cover that possibility, the company sent Hunsaker to San Francisco on 19 January with instructions to meet Rossby and discuss plans for a communication system to serve the Western Air Express route. After ten days in San Francisco, Hunsaker went on to Los Angeles, where he talked with the general manager of AT&T's affiliate, Pacific Telephone and Telegraph Company (PT&T), about ways the company could support an airway project.[20]

Back in New York, Hunsaker concluded that AT&T, with its demonstrably superior technology and the administrative and managerial expertise to make it work, had to present a more convincing case to the government and the airlines. As part of its mandate to provide aids to navigation, the Department of Commerce had installed a point-to-point airways radiotelegraph system and was ready to ask Congress for additional appropriations to upgrade its existing radio communications facilities. Increasingly, the Bell Labs–AT&T network of ground lines looked expensive and redundant. Hunsaker examined the government plans and found that they included unrealistic cost estimates and failed to take into account the frequency interference problems likely to arise when additional radiotelephones and radio beacons went into service on airways. AT&T's recently developed automatic teletype system using land lines offered distinct advantages for point-to-point communications, allowing both direct monetary savings and more room on the frequency bands for radiotelephone communication with aircraft. Additionally, he pointed out that Bell Labs' new radio system, which combined voice communication with a radio compass, was much superior to one proposed by General Electric for the Department of Commerce. It was frustrating to Hunsaker to have the government and the airlines look elsewhere when AT&T was ready with a comprehensive system that would integrate wire and wireless systems for airway communications, navigation, and weather reporting.[21]

Persistence paid off. In May 1928 Hunsaker reached an understanding with MacCracken and the Aeronautics Branch to evaluate the AT&T system for the New York–Chicago airway. Teletype machines leased from AT&T were to be installed at Hadley Field, Bellefonte, Cleveland, Bryan, and Chicago. Regular long-distance telephone lines would be used for meteorological information gathered from a secondary network of Weather Bureau stations located off the airway, with the data collected and dispatched immediately using the teletype system to all other stations on the route. At another station in the Commerce Department in Washington, people monitored operations and weather along the entire route. The department began working on the system during the summer and had the complete service in operation before the end of 1928.[22]

In the midst of determining AT&T's role in the New York–Chicago airway project, Hunsaker and his family moved from Brooklyn to Glen Head, Long Island, about twenty-five miles from Manhattan on the north shore. In November 1927 Hunsaker learned through Nelson Doubleday, a friend from his Aircraft Division days, that a 1750s two-story Dutch colonial farmhouse was for sale on about three acres on Cedar Swamp Road in Glen Head. He told a friend that the house was "old and ill used" and that it "needs a lot of restoration" but the price was fair. The Hunsakers quickly negotiated a deal with the owner. Over the winter, the Hunsakers brought in people to do the repair and renovation work, spending more than $10,000 to get the house in shape. Alice employed a prominent landscape architect to help her design and cultivate a formal garden similar to those she had seen in England. The result was spectacular—a tasteful mix of trees, shrubs, and flowers, set off by a flagstone terrace and white brick walls. Hunsaker lamented that he now had about an hour's commute into New York by train, but he and the family liked the ambience of the old house and the refreshing sense of space and freedom on Long Island.[23]

While the Hunsakers were getting the Glen Head house in shape to move in, Harry Guggenheim and Harris Hanshue finalized an agreement on a West Coast airway. From his research, Rossby concluded that there was enough variation in weather conditions on the Los Angeles–San Francisco route to provide a fair test of the system. Although California weather was usually good, pilots often encountered cloudy conditions in some of the mountainous regions, and dense morning fog occurred along the coast and in the central valley, where it sometimes persisted for a week at a time. Rossby envisaged the airline using timely forecasts in dispatching aircraft and directing them along one of several alternative routes between the two cities. The result would be more flexibility in airline operations, a significant increase in the number of flying days, and a boost in the company's revenue.[24]

Rossby's plan centered on a network of weather stations between Los Angeles and Oakland, each of which was to report three times daily to meteorologists in the two terminal cities, who drew up forecasts and advised departing pilots about conditions. Radio-equipped aircraft received updates en route. In March 1928 Guggenheim notified the Commerce Department and other government agencies that the fund would grant $60,000 to Western Air Express for a weather and communication system for the model airline, which provided passenger service over the 365-mile (587-kilometer) Los Angeles–San Francisco route.[25]

Although he preferred the New York–Chicago airway to that proposed for the West Coast, Hunsaker concluded that AT&T's involvement was critical if the company were to make a strong case for providing the equipment and services for a nationwide system. Hunsaker took the long rail journey to California at the end of February to touch base with people at PT&T and to see what needed to be done to modify the New York–Chicago proposal to mesh with the Guggenheim–Western Air Express plan. Harry Guggenheim met in San Francisco with PT&T representatives, who liked the idea of providing the service on the Los Angeles–San Francisco route. Unfortunately, Guggenheim wanted PT&T to make the service gratis for the year-long duration of the experiment, whereas the company was willing to donate only six weeks of service, after which it would charge commercial rates. With negotiations at another standstill, Guggenheim returned to New York greatly discouraged.[26]

It again fell on Hunsaker to try to break the stalemate. He met with Guggenheim shortly after his arrival back in New York and pointedly asked him why the fund would not pay for the communications service after the initial start-up period. Guggenheim answered that he thought that six weeks was not enough time to determine whether the system worked—that at least six months would be necessary. With a longer trial, more data could be obtained and a more persuasive case could be made to the airlines and the government that such a system could be extended to other airways. Finally, the Guggenheim Fund had been created to provide money for projects that neither the government nor private enterprise were in positions to support. This was obviously not the case, because AT&T had demonstrated that it had both the money and the facilities to back the experiment.[27]

Hunsaker was convinced. He discussed the matter in early May with one of AT&T's vice presidents and reached an agreement that Pacific Telephone and

Telegraph would not donate any services to the Guggenheim Fund but would instead function as an equal partner with the Guggenheim Fund and Western Air Express in an experiment for the government in the public interest. They further agreed that PT&T would provide its services gratis for at least a year in order to obtain "a conclusive result of value to us and to the Air Transport Companies." Hunsaker also received assurances from Bill MacCracken that his department would back the plan and would seek the full support of the Weather Bureau. The understanding was a big concession for PT&T, but it demonstrated Hunsaker's perception that a short-term investment by AT&T "should lead to a special commercial service that could be offered any Air Transport Company" while enhancing the image of the company to both the government and the general public.[28]

With this last piece of the puzzle in place, Rossby hurried to complete the weather reporting and communication system. At the heart of the network, when it began operations on 21 May 1928, were twenty-two reporting stations, located on both sides of the airway from Los Angeles through Bakersfield and Fresno to its northern terminus at Oakland. The scattered weather stations reported three times daily by telephone to the main Los Angeles and San Francisco offices, but they could, on demand, make reports every half hour. Each report included information on wind direction and speed, visibility, precipitation, cloud cover, temperature, and barometric pressure. Meteorologists at the northern and southern terminals compiled the data and issued forecasts, which pilots used in selecting one of five alternative routes for their flight plans. A two-way radiotelephone system provided weather updates to aircraft en route. The system did not, however, use teletype equipment for point-to-point communication.[29]

With appropriate ceremonies and ample attention from the press, Western Air Express inaugurated flights over the route on the morning of 26 May. Operations that day and over the next twelve months went smoothly, with the airline achieving a 99 percent on-time rate and an impeccable safety record. Passengers praised the airline for its performance and extolled the comfort and speed of the big Fokkers. The volume of traffic on the route, used not only by Western Air but also by Pacific Air Transport and Maddux Air Lines, climbed so rapidly that within a matter of months eighteen more reporting stations had to be added to the service, and forecasts were issued six times daily. Bakersfield and Fresno augmented Los Angeles and San Francisco as data collection centers, linked by telephone lines to the new stations.[30]

Hunsaker closely monitored the operations of both airways throughout the summer of 1928, paying particular attention to the route on the West Coast. He wrote in June that "we are all glad to hear that the . . . operation is going off so

Passengers boarding a Western Air Express F-10 Fokker trimotor, used on the West
Coast Model Airline in 1928–29. (A-2248, Courtesy NASM)

smoothly and that the service is proving useful. I can only hope they run into a
spell of bad weather before long so that the values of the service will be mag-
nified." Under Rossby's direction, the meteorological system ran well, garner-
ing compliments from both Western Air pilots and army aviators, who also relied on it for weather information and flight planning. Hunsaker was dismayed
by the Weather Bureau's apparent lack of interest in fulfilling its part of the bargain. By the middle of July, the bureau had failed to assign any of its personnel
to the weather service and had made no plans to provide money to continue the

operation after the initial trial year. Hunsaker surmised that part of the problem was lingering friction between Rossby and the Weather Bureau, which remained hostile to Rossby's methodology. In any case, Hunsaker wanted the PT&T people to keep track of Rossby's expenses so that when he met with Weather Bureau officials in Washington he could make the strongest possible case for the cost-effectiveness of the service. He also urged PT&T officials to encourage Rossby to use the company's teletype system for dispatching weather information from airport to airport.[31]

The airway communications project was a personal triumph for Hunsaker and a technical achievement for Bell Labs and AT&T. The integrated system worked so well on the New York–Chicago and Los Angeles–San Francisco routes that the commerce department and the Weather Bureau expanded it to include 8,000 miles (12,875 kilometers) of the nation's airways by the end of 1930. AT&T's automatic teletype machines and dedicated ground lines between major terminals were part of the reason for the success of the system, allowing the transmission of a high volume of data while freeing the airwaves for radio communication between ground stations and aircraft en route. The radio equipment developed by Bell Labs and manufactured by AT&T also proved its worth in service over the next few years, as the nation's air transport system underwent exponential growth.[32]

The Bell Labs airway communication project is a good example of the cooperative nature of airway development in the late 1920s, involving a complex mix of government and private interests, all of which had to be balanced against one another. There was often confusion and conflict, but in the end the nation got what was perhaps the world's most efficient airway communication system. Without this essential piece of infrastructure, the development of a comprehensive and efficient commercial aviation system would not have been possible.

Hunsaker's role was brief, but critical. He understood, perhaps better than anyone else in the company, the importance of aviation and its potential. AT&T wanted to maximize its role, but its corporate leaders had little vision of what and how big that role would be. Emphasizing a rational approach to the problem, Hunsaker combined his knowledge of engineering and aeronautics with his organizational experience to define the requirements, marshal the necessary resources, and put together a program to accomplish what he and the company wanted. Much like his work with the Aircraft Division in World War I, the airways project was more an administrative than a technical achievement, demonstrating once again Hunsaker's role as a heterogeneous engineer and setting the stage for the next phase of his career in the aviation business.

7.   Goodyear-Zeppelin

n March 1928 Paul W. Litchfield, now president of the Goodyear Tire and Rubber Company and the Goodyear-Zeppelin Corporation, approached Jerome Hunsaker in connection with his company's bid on a navy contract for the construction of two big rigid airships. Hunsaker and Litchfield had known each other for years. Both had MIT connections (Litchfield graduating with the class of 1896), they had worked together during the war on Goodyear's naval airship program, and they had both been with the American aviation mission in Europe in 1918–19. Despite Hunsaker's private remarks then about Litchfield's narrow outlook on the world, he respected Litchfield's business sense and experience in the lighter-than-air field. Over the next few months, the discussions between the two broadened to include the airship program generally and how it might serve as a springboard for a potentially lucrative commercial airship industry.[1]

Before Hunsaker could do much with Litchfield and Goodyear in the summer of 1928, he had to get his family settled at Wildwood, the rustic retreat they had bought in late 1926 at St. Huberts in the Adirondacks. The cool mountain air, spectacular scenery, and proximity to some of the best trout fishing in the East made the place particularly appealing to Hunsaker, who made a point of spending a couple of weeks a year at the place. Back in the city, he undertook at Litchfield's request preliminary studies of the economic potential of large rigid airships. At the time he was still with Bell Labs, and there was no indication that he was unhappy with his position. He had found AT&T's airways communication project satisfying as an exercise in applied research, but after assessing

Hunsaker and Paul Litchfield of Goodyear Tire and Rubber Corporation. (JCH 14, Courtesy MIT Museum)

his situation in the spring and summer of 1928 he did not see much more that Bell Labs could do in aviation. The airship, on the other hand, seemed poised to move from the experimental stage to commercial reality. When Litchfield offered him a job with Goodyear-Zeppelin in early August, Hunsaker did not hesitate to accept—especially when he understood that the company would keep him in New York and that he would not have to leave the Glen Head house.[2]

The Goodyear-Zeppelin Corporation had emerged from a collaboration between Goodyear Tire and Rubber and Germany's Luftschiffbau Zeppelin, which sought an American partner to help with commercial ventures. At the same time, Goodyear viewed a connection with the Zeppelin company as the most expeditious way of gaining access to the latest rigid airship technology. In 1921 Litchfield negotiated an agreement with Hugo Eckener of Luftschiffbau Zeppelin by which the new Goodyear-Zeppelin firm received the German airship company's rights and patents in North America, along with access to its most experienced personnel, in return for a two-thirds financial interest in the company.[3]

Behind the Goodyear-Luftschiffbau alliance lay the fundamental economic-technical assumption in the mid-1920s that the large rigid airship was the only aircraft capable of transoceanic flights carrying a payload of passengers and cargo. The airplanes of that era had steadily improved in reliability and performance, but with limited-capacity engines, inefficient propellers, and structures that had not changed much from before the war, they remained incapable of crossing oceans except under the most unusual circumstances. It was not simply a matter of making airplanes larger to increase their range and load-hauling capacity; despite advances in materials and construction, aeronautical engineers found themselves in an ever-tightening noose of weight and drag as they scaled up their designs. Unless there were a breakthrough in a combination of structures, power plants, and propellers, the airplane offered little hope for transoceanic operations.[4]

It was a different picture altogether with the rigid airship, for which lift was an independent variable rather than a direct function of aerodynamics and power plant performance. Rather than a liability, as it was in an airplane, size was a dividend in an airship: There was a direct correlation between the volume of lifting gas and range and payload capacity. Moreover, expanding the diameter of an airship's hull led to exponential increases in its volume while keeping the length of the craft within reasonable limits and avoiding excessive bending and shearing moments along its longitudinal axis. This is not to say that there were no drawbacks. Fabricating lightweight materials was a challenge, there were persistent structural design obstacles as airships grew larger, and there was the ever-present danger of fire in the use of hydrogen as a lifting gas. But most engineers thought that the materials and structures problems could be overcome and that nonflammable helium, though more expensive than hydrogen and lacking the lifting power of the lighter gas, was the answer to the safety issue. Finally, airships, unlike airplanes, could not land just anywhere; they needed large-scale and expensive base and support facilities.[5]

With the airplane out of the question for intercontinental travel, logic led to a comparison between the airship and the fast ocean liner. The assumption was that the airship fit into a niche. Though only half as fast as an airplane, the airship was three times faster than a crack transatlantic steamer, and it could approximate the liner in the comfort and style of its accommodations. There was every reason to expect—and most airship advocates stressed this—that the airship was ideally placed to skim off the cream of transoceanic travel and appeal to a small percentage of people willing to pay a premium for speed and luxury. Social and economic considerations reinforced what many thought to be the clear technological advantages of the airship.[6]

Critical to the success of the rigid airship was an industrial organization for the construction and operation of the huge and complex craft. In late 1924 Litchfield brought from Friedrichshafen a cadre of engineers, led by Karl Arnstein, as the nucleus of an airship design and engineering staff. Goodyear-Zeppelin was ready when the landmark 1926 five-year aircraft procurement legislation authorized the navy to acquire two 6-million-cubic-foot (169,900-cubic-meter) airships, although a year passed before Congress provided funds for a design competition. Goodyear-Zeppelin won, but not until October 1928 did the company sign a contract with the navy to build one airship of 6.5-million-cubic-foot (184,060-cubic-meter) capacity for $5,375,000, followed by a second of similar size costing $2,450,000—by far the most expensive aircraft the government bought between the wars.[7]

The navy was vital to the success of Goodyear-Zeppelin's plans for the commercial development of large rigid airships. Without the 1928 navy contract the company could not hope to create the infrastructure needed to design and build such complex craft, nor could it otherwise lure investors into such an expensive and risky undertaking. In turn, the navy recognized that it had to rely on a secure private industrial base for the future procurement of military airships. Admiral Moffett, in particular, took every opportunity to emphasize that American leadership in commercial airship development was crucial to economic prosperity and the nation's defense.[8]

Hunsaker told one of his British friends in August that he was "leaving the Laboratories to engage in a new venture or adventure, the building of Zeppelin Airships for the Navy and for Trans-Atlantic commercial service." As vice president in charge of commercial affairs, Hunsaker assumed his new position on 1 September and immediately launched into an examination of the feasibility of commercial airship operations. Knowing how important accurate weather forecasting and reporting was to safe and efficient commercial aviation, he completed a preliminary investigation of Atlantic weather conditions. He found that the northern, or Great Circle, route from west to east had the advantage of prevailing winds but that it also might subject an airship to some of the worst storms in the Atlantic. During such periods of bad weather, an airship was better making the west-to-east crossing on a more southerly course, near the Azores. As for east-to-west passages, the conventional wisdom was that a southern route was best, but Hunsaker's initial data showed that at some times the northerly Great Circle route was preferable.[9]

The Goodyear-Luftschiffbau relationship received a welcome boost with the spectacular transatlantic flight of the new German LZ.127, the *Graf Zeppelin,* in the fall of 1928. With five 550-horsepower Maybach engines, the 3.7-million-cubic-foot (104,770-cubic-meter) hydrogen-filled airship cruised at seventy-three miles (117 kilometers) per hour and had an endurance of up to 120 hours—more than enough to span the Atlantic carrying twenty passengers and forty crew members. The *Graf Zeppelin* flew for the first time in September 1928 and after only four more test flights weighed off from Friedrichshafen on 11 October bound for the United States under Eckener's command. After a stormy flight, the airship arrived at Lakehurst on the evening of 15 October.[10]

Hunsaker spent the next few days conferring with Eckener and others about a transatlantic airship line involving Goodyear-Zeppelin and the Luftschiffbau. Following a round of meetings in New York on the seventeenth, Eckener, Hunsaker, and Karl Arnstein went by train on the morning of 18 October to Lakehurst, where they toured the big hangar and inspected the *Graf Zeppelin.* At lunch the discussions focused on the financing needed to get commercial operations under way. Hunsaker left for New York, while Eckener went to Washington to meet with Hunsaker's friend, Edward P. Warner, formerly a professor of aeronautical engineering at MIT and now assistant secretary of the navy for aeronautics. Chicago was next on Eckener's tour, followed by a meeting with Litchfield in Akron before returning to New York. The *Graf Zeppelin* lifted off from Lakehurst early in the morning of 29 October and landed at Friedrichshafen seventy-two hours later, on 31 October.[11]

Determined to translate the public enthusiasm generated by the *Graf Zeppelin* flight into financial support for commercial airship lines, Litchfield and Hunsaker continued informal talks with Wall Street bankers and investment houses. Hunsaker prepared a proposal based on the supposition that the airship had "arrived at the point where its application for commercial overseas transportation is fully justified." Included in the proposal were descriptions of three passenger airships based on the design of those being built by Goodyear-Zeppelin for the navy.

The first airship accommodated a hundred passengers on two decks within the hull, with amenities that included elegant art deco staterooms, a dining salon, lounge, ballroom, smoking room (complete with ersatz fireplace), library, and confectioner's stand. A promenade off the upper deck allowed travelers to sit or stroll about while viewing the scenery passing below. The second airship carried forty passengers with a much larger mail and express payload and a greater fuel load for longer range. In the third airship, seventy-five passengers could be carried with about half the weight of mail and express as the second alternative. Hunsaker was confident that "the American public, educated as it is to the best obtainable in transportation, will not be hesitant in accepting comfortable and luxurious airships" for safe and efficient travel across the oceans.[12]

It was understood that a government subsidy was crucial to the success of the Goodyear-Zeppelin airship scheme. Litchfield had long been a staunch backer of the ruling Republican party; so close were he and his company to the party that people joked that GT&R stood not for Goodyear Tire and Rubber but for "Generally True Republicans." He also knew Herbert Hoover personally. Litchfield and Hunsaker met with the president on 9 May 1929. They outlined Goodyear-Zeppelin's proposal, committing the company to a $10 million investment but emphasizing that unless there was full support from the federal government it was unlikely that the company would be able to lure the capital needed for the enterprise. Absolutely necessary was an airmail contract, preferably for three years—long enough, Litchfield and Hunsaker thought, to build up passenger revenue. If Congress authorized the Post Office to enter into airmail contracts before the year was out, Litchfield estimated that operations could begin as early as 1932. Hoover ended the meeting with a promise of full White House support for the scheme.[13]

The *Graf Zeppelin*'s 1928 transatlantic crossing was sensational enough, but it paled in comparison to the airship's round-the-world flight in 1929. To be eligible for $100,000 offered by William Randolph Hearst, the *Graf Zeppelin* had to begin and end its flight in New York City. So the first leg of the journey was a dash across the Atlantic to Lakehurst and an overflight of Manhattan before the airship returned to Friedrichshafen to prepare in earnest for the global voyage. Leaving Germany on 15 August, the *Graf Zeppelin* flew east over Russia to Japan, spanned the Pacific to San Francisco and Los Angeles in a little more than seventy-nine hours, and crossed the country to New York, before ending the epic flight at Lakehurst on 29 August. After relinquishing command of the *Graf Zeppelin* to Captain Ernst A. Lehmann for the return flight to Friedrichshafen on 1 September, Eckener used the rest of his time in the United States to firm up plans with Goodyear-Zeppelin for a commercial airship service.[14]

A series of meetings with Litchfield and Hunsaker resulted in an understanding that Goodyear-Zeppelin and the Luftschiffbau would cooperate in the creation of airship lines across both the Atlantic and the Pacific. The transatlantic enterprise would be a partnership between the American and German firms, whereas the transpacific route would be served by a completely American-owned company. Four airships would be built—two by Goodyear-Zeppelin, for use in the Pacific, and two by the German company, for the Atlantic route. The Americans promised to build a large hangar and support facility somewhere on the East Coast, supplemented by mooring masts in New York City and Philadelphia. National City Bank and the investment houses of Lehman Brothers and Grayson M.- P. Murphy would put together the estimated $15 million financial package. All parties recognized that the viability of the enterprise ultimately rested on a substantial government subsidy in the form of an overseas airmail contract.[15]

Following up on these discussions, Eckener met in Berlin with Charles E. Mitchell, chairman of the board of the National City Bank. The two agreed to organize the International Zeppelin Transport Company, which officially came into existence on 18 October. Primarily a "paper corporation," and capitalized at only $100,000, International Zeppelin was to undertake a year-long study of the feasibility of transatlantic airship operations and to explore means of financing them should such operations merit consideration. It was understood that if the studies showed the practicality of transatlantic service, the company would spin off one or more operating companies as subsidiaries, with Goodyear-Zeppelin and the Luftschiffbau retaining their fifty-fifty partnership.[16]

Subsequent talks between Mitchell and Eckener in March 1930 hammered out the details of the organization of International Zeppelin. Considering the uncertain economic climate in the wake of the October 1929 stock market crash, the growing list of firms willing to be involved in what might well turn out to be a high-risk enterprise spoke to the continuing enthusiasm for transoceanic air travel. In addition to Goodyear-Zeppelin and its German partner, three other companies backed International Zeppelin: United Aircraft and Transport Corporation, one of the nation's largest aviation holding companies; Union Carbide and Chemical Company, which among many activities was engaged in the extraction of helium; and Alcoa, long involved in the development of duralumin for aircraft. The National City Bank joined Lehman Brothers and Grayson M.-P. Murphy in coordinating the financial arrangements for the new company. Its board of directors reflected the confidence of both

Wall Street and the aviation community in the potential of commercial airship operations. Serving as chairman of the board was United Aircraft's chairman, Col. Edward A. Deeds. Litchfield was its president, Eckener one of the directors, and Hunsaker was vice president and general manager. Other prominent members of the board were Roy A. Hunt, president of Alcoa, Frederick B. Rentschler, president of United Aircraft, and F. W. ("Willy") von Meister, the American representative of Germany's Maybach Motors company, a supplier of airship engines.[17]

Concurrently with the creation of International Zeppelin, Eckener and Litchfield reached an understanding on the formation of a similar company to investigate transpacific airship operations. Chartered four days after International Zeppelin, the Pacific Zeppelin Transport Company had Hunsaker as its president, Litchfield as chairman of the board, and Edward P. Farley of the American-Hawaiian Steamship Company as chairman of the executive committee. Directors included the requisite aviation people—Rentschler of United Aircraft, Clement M. Keys of Transcontinental Air Transport, Graham B. Grosvenor of the Aviation Corporation, Juan Trippe of Pan Am, and James A. Talbot of Western Air Express, as well as a liberal sprinkling of executives associated with ocean shipping on the West Coast. With half the capital of International Zeppelin, Pacific Zeppelin was another paper corporation, charged with studying the potential of airship operations from the West Coast to Hawaii and the Philippines.[18]

Another element of Goodyear-Zeppelin's plans was the Daniel Guggenheim Airship Institute in Akron, an outgrowth of Hunsaker's desire to bring basic research to the lighter-than-air field and to provide a firm foundation for airship development. In February 1929 he called for the creation of an institute where "a few advanced thinkers might be put to work with benefit to the art," later approaching Harry Guggenheim to see if his fund had any interest in underwriting the concept. Guggenheim liked the idea and on June 12 invited Hunsaker and his friend Robert A. Millikan, a physicist and the president of the California Institute of Technology, to his estate on Long Island to discuss the proposal. The meeting resulted in an agreement whereby the Guggenheim Fund would contribute $250,000 to establish the institute in Akron, provided the city matched the grant with enough money to cover operations for a five-year period. Working with Hunsaker, George F. Zook, president of the University of Akron, put together a package that included the donation of a tract near the municipal airport and a bond issue to cover estimated operational expenses.[19]

To foster support in southern California for Goodyear-Zeppelin's proposed transpacific airship service, Hunsaker insisted that the Guggenheim Aeronautical Laboratory of the California Institute of Technology administer the institute and that fellowships in Caltech's Aeronautical Engineering Department be set aside for lighter-than-air research. He and Millikan, with Harry Guggenheim's approval, devised a complicated arrangement whereby the Guggenheim Aeronautical Laboratory received money from the Guggenheim Fund to cover the costs of administration, faculty and staff salaries, and fellowships associated with the airship institute over five years.[20]

The linchpin of the plan was attracting a recognized figure in aerodynamic research to lend prestige to the institute and to provide the necessary direction and leadership during its formative years. Hunsaker knew who he wanted—Theodore von Kármán, director of the Aerodynamics Institute at Aachen, who had worked under Prandtl at Göttingen and had done wind tunnel investigations on the ZR-3 for Luftschiffbau Zeppelin. Millikan liked the idea, too. He had ambitions to see Caltech develop along similar lines to the University of Göttingen and had brought von Kármán to Caltech in 1926 as a consultant to help with the design of a closed-circuit wind tunnel. Hunsaker was certain that a joint position as head of the Aeronautical Engineering Department at Caltech and director of the new airship institute would be sufficient inducement to lure von Kármán back to the United States. After receiving assurances from Harry Guggenheim that the airship institute directorship would not prevent him from continuing his research at Caltech, von Kármán accepted the offer, but not until 20 October. The day before, with only the vaguest assurances that von Kármán was safely on board, Harry Guggenheim announced the creation of the airship institute, followed the next day by Goodyear-Zeppelin press releases on the formation of the Pacific Zeppelin and International Zeppelin commercial airship enterprises.[21]

Hunsaker, Millikan, and Guggenheim were among the hundreds present at the dedication of the airship institute on 26 June 1932. Housed in a four-story building at the Akron airport, the institute had ample space for classrooms, laboratories, and offices. The crown jewel, however, was a six-and-a-half-foot (two-meter) -diameter vertical wind tunnel, which generated speeds up to 120 miles (193 kilometers) per hour. The idea was to cancel out lift as a variable by hanging airship models vertically, permitting more accurate measurements of drag and turbulence, both major concerns with large rigid airships. A smaller wind tunnel, used primarily for boundary-layer research, was also located in the building.[22] An example of Hunsaker's continuing commitment to basic research in aerodynamics and his skills in bringing together potentially competing forces to achieve his goals, the institute was a superbly equipped and generously funded research facility and helped stimulate Goodyear-Zeppelin's long-range plans for the development of commercial airships.

Hunsaker on the cover of *U.S. Air Services* magazine in the spring of 1932, when he was vice president of Goodyear-Zeppelin Corporation and deeply involved in international airship transportation. (Courtesy NASM)

In large part, the viability of Goodyear-Zeppelin's commercial airship program centered on the establishment of a new operating base and terminal on the East Coast. In late March and early April 1930, Hunsaker, Eckener, and others involved with the project visited sites from Philadelphia to Richmond. The Hybla Valley tract south of Alexandria, Virginia, then being used as a landing field for student aviators, seemed best, but so did other sites near Moorestown, New Jersey, and Linthicum, Maryland. Their plans for an airship base came amid a veritable airship boom in 1930. The *Graf Zeppelin* followed its round-the-world odyssey with an equally sensational "triangle flight" in late May and early June, from Europe to South America, north to Lakehurst, and back to Germany. On 1 August the 5-million-cubic-foot (141,584-cubic- meter) British airship R.100 crossed the Atlantic to Montreal, carrying a handful of passengers, including the aeronautical engineer and later best-selling novelist Nevil Shute. Hunsaker inspected the airship and the mooring facilities that had been built for it at St. Hubert, south of the city. In anticipation of regular intercity airship travel, the architects of the Empire State Building in Manhattan included a 205-foot (62.5-meter) mooring mast, which Hunsaker, Eckener, and von Meister examined in late March 1931, a month before the building opened.[23]

From the beginning of the enterprise it was understood that there should be a free exchange of information between the Germans and the Americans. Not least among those unhappy with this arrangement were many German engineers, who believed the flow of technical information benefited only the Americans. Later, as Eckener began planning for a larger airship in anticipation of regular transatlantic service, the Germans became grateful recipients of American expertise. Following the *Graf Zeppelin*'s spectacular global flight in the summer of 1929, the Luftschiffbau Zeppelin began preliminary design work on its successor, the LZ.128, a 5.3-million-cubic-foot (150,080-cubic-meter), hydrogen-inflated airship able to accommodate up to thirty-four passengers in transatlantic service. All of the design work on the LZ.128 had been completed and one ring of the framework had been laid by the fall of 1930, when the world learned of the loss of the new 5-million-cubic-foot (141,584-cubic-meter) British airship, the R.101. Encountering a vicious storm, the airship crashed on the night of 4–5 October near Beauvais, France, its hydrogen gas igniting and causing a conflagration that killed forty-eight passengers and crew. Horrified by the accident, which cast a pall over the intimate airship community, Hunsaker told the press that "it is apparent that if helium had been used in the R.101 instead of hydrogen a great loss of life would have been prevented." Litchfield agreed and after meeting with Hunsaker dispatched him to Friedrichshafen to talk to Eckener about the LZ.128 project.[24]

Mast and terminal at the new Empire State Building, Manhattan, 1931. No airship ever moored there. (88-6716, Courtesy NASM)

Hunsaker met with Eckener in late October to discuss modifying the LZ.128 to use helium and to go over the partners' plans for a transatlantic passenger operation. Rather than undertake a redesign, Eckener decided to build an entirely new airship, inflated with helium and designated the LZ.129. The ship would have a capacity of 6.5 million to 7.5 million cubic feet (184,060 to 212,376 cubic meters) and carry about forty passengers and a large quantity of freight. For his part, Hunsaker promised that International Zeppelin would do everything it could to ensure German access to American helium. The negotiations included an understanding that both the Germans and the Americans would construct new airship terminals in 1931, with the new German airship beginning transatlantic service that year and the first American airship starting a year later.[25]

From Germany Hunsaker journeyed to Paris for the first meeting of the International Aerial Safety Congress in December. He made "a very sad trip" to Beauvais, where the wreckage of R.101 reminded him of the human cost of airship pioneering and reinforced his commitment to helium. At the International Congress he read a paper on the safety of rigid airships that in effect was "damage control" in the wake of the R.101 accident. In the presentation Hunsaker stressed that airship crashes to that point were attributable either to poor design or to lack of experience in their operation. "Modern airships," he maintained, "should be as little subject to structural failure as well-constructed steamships." He went on to explain the "triple-layer principle" in the design of rigid airships, incorporating a "highly redundant system" of structural components, separate gas cells (analogous to a ship's watertight compartments), and a fabric covering that afforded protection from the elements. Engine failure rarely occurred; when it did, the airship could still maintain headway with half its power plants. Finally, accurate weather forecasting and the latest navigation equipment assured the widest possible margin of safety in passenger operations.[26]

Back home before Christmas, Hunsaker turned his attention to obtaining a federal subsidy for Goodyear-Zeppelin's commercial airship ventures. From the 1925 Kelly Act on, the government's support had come in the form of long-term airmail contracts to private operators, with payment first on a weight-per-mile formula and later, following new legislation in 1930, on a space-per-mile basis. The subsidy was in the loss the government incurred between the price of airmail postage and the payments to the carriers. Acceptable for airplanes with their limited capacity, the space-per-mile provisions if extended to airships would result in astronomical government outlays.

In late 1929 Hunsaker and Pacific Zeppelin's Edward Farley had spoken to California Republican Senator Hiram W. Johnson about the possibility of securing a federal subsidy for commercial airship operations. Johnson persuaded

one of his colleagues, Charles L. McNary, a Republican from Oregon, to spear-
head the legislation on Capitol Hill. On 16 April 1930 McNary and Congressman James S. Parker, a Republican from New York, introduced identical airship
bills. Drafted with the assistance of Bill MacCracken, who had left the Department of Commerce and joined a Washington law firm, the two measures, once
consolidated, became known as the Merchants' Airship Bill.[27]

The proposed legislation resembled laws providing federal aid to the country's merchant marine, which encouraged the construction and operation of
cargo vessels and passenger ships deemed vital to the nation's economy and
security. Companies operating airships in foreign commerce would be eligible
for mail payments up to twenty dollars per mile provided their equipment was
capable of carrying no less than 10,000 pounds (4,536 kilograms) and had a
minimum range of 2,000 miles (3,219 kilometers) without refueling. Finally,
the bill authorized the creation of a special fund, intended to provide a loan
of up to ten years to cover 75 percent of airship construction costs. As the bill
ground its way through the legislative mill, Litchfield and other advocates of
the airship exerted quiet pressure on lawmakers. Nevertheless, when the bill
was reintroduced in December, after several redrafts in consultation with the
war, navy, and commerce departments, it no longer included the loan and specified that the airship companies must bear the full burden of construction
costs.[28]

Hunsaker spent the last few days of December 1930 and most of January
1931 preparing for congressional hearings on the McNary-Parker bill. Central
to Hunsaker's work was an intriguing "war game" that he and his staff played.
Using five years of meteorological information obtained from the Weather Bureau, Hunsaker's group confirmed that bad weather made the Great Circle route
across the north Atlantic totally unsafe for wintertime use; the route farther to
the south, through Bermuda and the Azores, was an ideal alternative. Hunsaker's staff then began a series of theoretical transatlantic airship flights. Starting with a weekly schedule using two airships—one flight originating in Paris
and the other in Washington every Saturday night—they plotted their courses
and progress across the Atlantic, making 520 "flights" over a five-year period.
Keeping to a sixty-hour average flight time in the summer and sixty-eight hours
in the winter, the airships could be expected to arrive on or ahead of schedule
75 percent of the time.[29]

Hunsaker's group carried out a similar study of routes for Pacific Zeppelin.
Again relying on years of weather data, they found conditions in the Pacific
good enough to maintain a year-round schedule. At the beginning, two airships
would fly once a week, making six-day runs from the West Coast to the Philippines. Westbound to Hawaii, airships could expect following winds 80 percent

of the time, cutting flight time to about thirty-six hours. Headwinds would extend the time of eastbound flights to forty-eight hours. From Hawaii to Guam there were also favorable winds, but from Guam to Manila and Tokyo the weather was more uncertain due to the prevalence of typhoons. Although these fierce storms were a potential threat, Hunsaker was confident that they could be easily predicted and avoided. Return flights from Tokyo to Manila were likely to meet crosswinds and headwinds, adding another day and a half to the journey. Hangars and major repair facilities would be located at either San Francisco or Los Angeles and at Manila, with mooring masts only at Honolulu, Guam, and Tokyo.[30] Rational and systematic, typical of Hunsaker's thoroughness, the airship studies were a superb rehearsal for the forthcoming congressional hearings.

The House Committee on Interstate and Foreign Commerce, chaired by Parker, began two days of hearings on the Merchants' Airship Bill on 28 January 1931. Hunsaker handled the technical aspects of Goodyear's case, beginning with a description of the type of airship Goodyear planned to use in commercial service. He explained that the smallest craft capable of regular transatlantic operations would be larger than the airship Goodyear was then completing for the navy—about 7.5 million cubic feet (212,376 cubic meters) in capacity, compared to the navy ship's 6.5 million cubic feet (184,060 cubic meters)—and more costly at $4 million to $5 million. The 850-foot (259-meter) -long commercial airship would accommodate eighty passengers in forty staterooms, require fifty crew members, and be able to carry 25,000 pounds (11,340 kilograms) of mail and express. Relying on drawings, maps, and charts, Hunsaker provided details of Goodyear's planned commercial airship operations. Studies of East Coast weather patterns showed that locations on the coast had more wind than those located farther inland; Lakehurst, the navy's principal airship operating base, was particularly bad, suffering from winds exceeding twenty miles (thirty-two kilometers) per hour for more than half the year. In comparison, Richmond, Virginia, had only 210 hours of wind annually. Hunsaker also went over the studies he had made of transatlantic weather and the theoretical airship operations that had been made based on the accumulated meteorological data.[31]

Hunsaker had done his homework on the economic side of the planned enterprise, too. Total costs, including the construction and operation of two airships, an East Coast terminal at $3.25 million, helium, spare parts, and working capital, came to about $16 million, which worked out to about $20 per mile. He went on: "Our potential market might be assumed to be half of all the first-class passengers who patronize the nine super-steamships and pay extra to do it." It was reasonable to suppose that half the 100,000 passengers who crossed the Atlantic each summer in fast liners might decide to make the journey by air. Of those 50,000 potential customers, half of them—that is, 25,000—paid $500 or more one way. It was a "pure guess," but one could conclude that those 25,000 people would be willing to pay twice as much—$1,000—to make the crossing in less than half the time of the speediest liner. Later, as the volume of traffic went up, the fare might be reduced to $750. As the service began, the passenger revenue from a typical crossing would be about $15,000, which, when added to the maximum mail rate specified in the bill ($20 per mile), gave a 6 percent profit on each trip. Each airship could be expected to have a service life of five years and would be depreciated to zero at an average of 20 percent per year. Over time, the number of passengers would increase and the mail pay would decrease, so passengers were essential to the economic success of the project; without them, Hunsaker said, "we are foolish to be wasting our time" with commercial airships.[32]

There was some urgency in getting the proposed airship service going. The Germans had been "captured by the romance of the Zeppelin development. It is Germany's favorite expression of national power. The Zeppelin will carry the flag of the New German Republic all over the world." If Congress acted immediately, the Germans would still have a substantial lead in the industry and the United States would have difficulty catching up; any delays meant the country would fall further behind. Already in place was the equivalent of a commercial agreement with Germany, in which the two countries shared the north Atlantic air routes on a fifty-fifty basis, but the Goodyear-Luftschiffbau Zeppelin arrangement expired in a few months, and with it would lapse any certainty of American involvement as an equal partner in the enterprise. Congress had to act quickly or run the risk of the country missing out on expanding foreign trade and a new transportation technology.[33]

On 6 February 1931, the Senate Committee on Commerce, chaired by Johnson, opened its own hearings on the Merchants' Airship Bill. Not surprisingly considering the constituency of the committee's chairman, much of the inquiry concentrated on Pacific Zeppelin's transpacific airship service. As he had done before the House committee, Hunsaker presented the results of the theoretical transpacific study, assuring the lawmakers that because of the generally advantageous weather conditions in the Pacific, the company could run a profitable year-round operation. "In the Pacific," he said, "we are fortunate in that the run from California to Hawaii is about the most favorable flying route anywhere in the world for airships," with moderate winds and temperatures. West of Hawaii, there were potential problems, especially seasonal typhoons, but he was sure that airships could be routed around the storms or held at Guam for better weather. With the *Graf Zeppelin* in 1929 Eckener had demonstrated how a knowledgeable airship commander could use Pacific weather patterns to make swift and safe west-to-east crossings.[34]

Operating two eighty-passenger airships, with terminals in California and the Philippines and mooring masts in Guam and Japan, the Pacific service would cost about $18 million—or $18 per nautical mile. Hunsaker did not think there would be many passengers or much express, especially west of Hawaii, which meant that it would be necessary to ask for the entire $20 subvention from the Post Office. As the service matured and attracted more paying customers, the subsidy might come down to $10 per mile. To the skeptics who thought the proposed service would compete with established shipping lines, Hunsaker assured them that a number of West Coast shipowners were directly involved in Pacific Zeppelin and that the company's service would complement rather than compete with existing transpacific lines.[35]

A bigger and totally unanticipated obstacle arose from Juan Trippe of Pan Am and Fred Coburn of the Aviation Corporation, one of the big aeronautical holding companies. Both insisted that the minimum amount of mail provided for in the airship bill be reduced from 10,000 pounds (4,536 kilograms) to 5,000 pounds (2,268 kilograms) and that airplanes be eligible for the subsidy. Because Trippe was a Pacific Zeppelin director and had not expressed any reservations about the airship bill until then, Hunsaker hastily arranged a meeting with Trippe and Coburn immediately after the Senate hearings to try to sort things out. Coburn explained that the Aviation Corporation was anticipating a big flying boat contract and needed government assistance. Trippe was almost hostile, alleging that Goodyear-Zeppelin had used its political influence to stymie his efforts to negotiate a transatlantic airmail contract for Pan Am. The talks continued the next morning, with Hunsaker agreeing to an amendment specifying the lower minimum poundage, then abruptly ended when Trippe and Coburn left for New York "to consult with their associates."[36]

Further negotiations failed to bring the two parties together. While Trippe met with Goodyear-Zeppelin lawyers in New York, Pan Am and Aviation Corporation agents lobbied in Washington against the airship bill. At the same time, Hunsaker met with Johnson in an effort to secure his committee's support for the legislation with the suggested changes to the mail poundage, only to find that the duplicitous Trippe and his allies wanted another round of hearings on the measure. With no settlement possible and time running out on the legislative session, Johnson decided not to consider the bill until all interested parties could agree on a compromise.[37]

The Seventy-second Congress convened in December 1931 with a Democratic majority in the House. Congressman Robert Crosser, a Democrat from Cleveland, introduced a new airship bill, unchanged from the old McNary-

Parker measure in that it excluded any reference to airplanes but included a pro-
vision delaying subsidy payments until three years after the law's enactment.
With the nation deeply mired in the Great Depression and antipathy widespread
to any legislation calling for additional federal outlays, the House Interstate
and Foreign Commerce Committee, now chaired by Democrat Sam Rayburn
of Texas, was lukewarm to the measure. Crosser persisted, and the committee
held perfunctory hearings in March, finally reporting the bill out on 3 May.
Goodyear launched a major lobbying drive, dispatching one of its blimps to
Washington to give complimentary rides around the capital to congressmen,
their staff, families, and friends. The effort paid off. When the floor debate
opened, most congressmen spoke in favor of the bill, although several opposed
it as nothing more than a thinly veiled subsidy to Goodyear-Zeppelin during a
time of economic crisis and pinched federal budgets. The House passed the bill
by a close roll call vote on 16 June.[38]

Hunsaker and the Goodyear-Zeppelin people could not rest on their laurels
for long, because Trippe and Pan Am exerted their considerable influence in
the Senate to have the McNary bill rewritten to reflect their interests. After several meetings, Hunsaker and Trippe agreed that Pan Am would not oppose the
Crosser airship bill and Goodyear-Zeppelin would support passage of the Mc-
Nary bill, with amendments extending the airmail subsidy to flying boats. The
Senate Commerce Committee reported out the McNary bill, as amended, and
on 23 June approved the Crosser bill with changes in wording consistent with
the Senate measure. In spite of intensive efforts to line up votes on the floor,
there remained a strong bloc of senators opposed to any airmail subsidies; the
compromise bill never reached a vote before July, when Congress adjourned
and most lawmakers went home to campaign for the November elections.[39]

Though discouraged by the failure of the airship bill for a second time, Johnson and other supporters of the measure remained optimistic about its prospects
when Congress reconvened in December. In the meantime, Hunsaker and
Goodyear-Zeppelin were busy with projects in hand, the most important being
construction at Akron of the first of the two big navy airships, designated the
ZRS-4. The work begun in Goodyear's huge new hangar, or "airship dock," in
November 1929 under Karl Arnstein's supervision proceeded through 1930
and into 1931. In many ways a prototype of Goodyear-Zeppelin's commercial
airships, the ZRS-4 was 785 feet (239 meters) long and had a gas capacity of 6.85
million cubic feet (193,970 cubic meters), up from the 6.5 million (184,060)
originally called for when the navy and Goodyear-Zeppelin signed the contract
to build this airship and its sister. Aft of the control car, tucked up in the hull
where the commercial airship would feature a large passenger cabin, the ZRS-4
had a hangar that accommodated five fighter planes.[40]

Exactly what role Hunsaker played in the building of the ZRS-4 is not easy to determine. He had, as he said, responsibility for the "general supervision" of the project, but working in New York and spending much of his time in Washington he would not have been able to oversee much, if anything, in Akron. Nevertheless, he routinely made trips to Akron in 1930 and 1931 to see how the airship was coming along, and Arnstein and others did not hesitate to consult him as the big airship passed various construction milestones. Named the *Akron* in honor of the city of its conception and birth, the airship was the focal point of week-long festivities, culminating in a christening by Mrs. Herbert Hoover on 8 August 1931. Hunsaker, Litchfield, Moffett, and others were among those present, along with hundreds more dignitaries and representatives of the press and the radio networks. Hunsaker was aboard for the *Akron*'s overnight delivery flight to Lakehurst on 21–22 October.[41]

There were other positive developments despite the legislative impasse in 1932. Even before the completion of the *Akron,* work got under way on its sister ship, the ZRS-5, named the *Macon* for the largest city in the Georgia congressional district of Carl Vinson, Democratic chairman of the House Naval Affairs Committee. Goodyear-Zeppelin also settled on Hybla Valley as the site for an East Coast terminal. By 1933, as the *Macon* neared completion, the

The big navy airship *Akron*, launched in 1931 and considered a prototype of the commercial airship. It went down in a storm off the coast of New Jersey in April 1933. (88-8008, Courtesy NASM)

company began fabricating subassemblies in Akron for the GZ-3, the first of its commercial airships. The Germans, meanwhile, continued work on the LZ.129, bigger still than the giant navy airships, with the expectation that it would be ready for transatlantic operations sometime in 1933.[42]

When the Seventy-second Congress reconvened in December 1932 for a lame duck session following the November elections, which had swept the Democrats to power in both houses and brought Franklin D. Roosevelt to the White House, optimism ran high that the McNary-Crosser bill would become law. The Senate seemed ready to pass the bill, and Hunsaker had lobbied those close to the president elect, gaining their assurances that Roosevelt supported the measure in part as a means of stimulating employment. Then on 2 March 1933, the day the Senate was to vote on the bill, Democratic Senator Thomas J. Walsh of Montana, Roosevelt's choice for attorney general, died of a heart attack on his way back from a honeymoon in Cuba with his young bride. To honor him, the Senate adjourned for the day, and when it reconvened on 3 March it neglected to revise the calendar. A day later, on 4 March, the session ended with the Merchants' Airship Bill left in a state of suspended animation.[43]

Hunsaker and others at Goodyear-Zeppelin still had every reason to believe the new Democratic Congress and the Roosevelt administration would get around to the airship bill, if not during the hectic first hundred days, then shortly thereafter. On Saturday, 1 April, Hunsaker prepared a report on the situation and mailed it to Litchfield in Akron on Sunday. The next evening, Monday, 3 April, the *Akron* weighed off from Lakehurst surrounded by rain and fog so thick that it had to cancel plans to take aboard its fighter unit. Rear Admiral Moffett and an official from International Zeppelin, Alfred F. Masury, were aboard for what was to have been a routine flight to calibrate the navy's new radio direction finding transmitters at Newport, Rhode Island. Five hours into the flight, while trying to skirt around or to penetrate a cold front preceding an immense low pressure center, the big airship crashed into the frigid waters of the Atlantic off the New Jersey coast. There were only three survivors. Moffett, the ship's commanding officer, Comdr. Frank C. McCord, and seventy-two others died in what was then the world's worst air disaster.[44]

Once the initial shock of the airship's loss passed, the navy moved with alacrity to form a court of inquiry headed by Rear Adm. Henry Butler to investigate the crash. The court convened at Lakehurst on 10 April and heard from a succession of witnesses, among them Lt. Comdr. Henry V. Wiley, the only officer to survive the accident, as well as those responsible for the design, construction, and operation of the airship. Of major concern, given the nature of previous airship accidents—and testimony from Boatswain's Mate Richard E. Deal, one of the survivors, that a major longitudinal girder had "opened" before the airship went down—was that a catastrophic structural failure had led to the *Akron*'s demise.

Appearing before the court on 19 April, Hunsaker insisted that the airship had been designed and built to incorporate a "higher order" of strength than any craft up to that time. Following the loss of the *Shenandoah,* he and other engineers had made extensive studies of airship structures, producing reams of data used in the strength calculations for the *Akron.* In severe gust conditions, he said, the *Akron* had a factor of safety more than two times higher than that of the German-built *Los Angeles,* and the aft portion of the hull—a weak point on earlier airships—could withstand aerodynamic forces far greater than those that the *Akron* had encountered on the night of 3–4 April. As for Deal's account, Hunsaker doubted that any of the longitudinal girders had broken in flight and wondered if it were even possible for Deal to have seen the girders in question from where he had been stationed.

On the other hand, Hunsaker surmised, if the lower fin had struck the water while the airship was under way and inclined upward at a sharp angle, extraordinary longitudinal forces might well have caused the suspect girders to bend or break—but under compression and not tension, as Deal's account seemed to indicate. It seemed to Hunsaker that the cause of the accident was a severe "downdraft," or wind shear, which rapidly dropped the *Akron* from its cruising altitude, and that in the attempt to climb back to safety a portion of the airship's lower fin and rudder struck the water. That in turn led to loss of control, and subsequent gusts drove the craft into the sea.[45]

The court of inquiry wrapped up its investigation on 27 April, concluding that it could ascertain no single reason for the accident. It was only "conjecture" that the *Akron* had flown into the sea that night, as Hunsaker hypothesized, or that there might have been a major structural failure, as Deal's testimony seemed to imply. There had been no "negligence or culpability," although the board found that McCord, in not taking a course to carry the airship around the storm, may have been partly to blame for the accident. Changes in procedures or course alterations that fatal night might have prevented the tragedy, but in the final analysis the loss of the *Akron* was due to a combination of factors, none of which by itself led to the disaster.[46]

Congress had its say in the matter, too, using the crash of the *Akron* as an excuse to examine the entire American airship program. A joint committee, chaired by Senator William H. King, Democrat from Utah, began hearings on 22 May. Hunsaker testified a week later. He repeated that the *Akron* was far and away the strongest airship ever built. The accident, to his mind, had been due to operational, not structural, problems; put simply, the *Akron* had been flying too low. He pointed out that the aneroid barometer used as an altimeter on the airship was not accurate enough to allow safe flying at minimal altitudes in poor visibility. Turning to broader policy issues, Hunsaker vigorously defended the navy's decision to build big rigid airships. In the economic climate of the times, and given treaty restrictions on the number and size of cruisers, the airship, he believed, continued to be a logical solution to the problem of providing adequate scouting and reconnaissance capabilities in the Pacific.[47]

Despite the crash and the appalling loss of life, Hunsaker remained convinced that it was too soon to give up on the rigid airship and called for a "continuing policy which will carry through the development period until such time as the future looks more clear." He considered it best to have the *Macon* operate with the Pacific fleet on the West Coast, where the weather was "more predictable" than it was in the Atlantic. Meanwhile, at Lakehurst, the nine-year-old *Los Angeles,* which had been laid up as an economy measure, should be recommissioned and made the center of a "training and research unit" aimed at "the improvement of the art" of airship operations. Specific attention had to be given to better weather forecasting, advanced instruments, and the instruction of personnel. Because of its age, the *Los Angeles* would have to be replaced by a more modern craft, the construction of which would help preserve and maintain the industry in the United States and guarantee that the navy had a modern training airship capable of operating in all weather. In keeping with his concept for an airship fleet large enough to sustain the inevitable operational losses, he urged the navy to build a replacement for the *Akron* as soon as possible.[48]

When all indications were that, in the aftermath of the *Akron* disaster, it would be difficult if not impossible to secure additional federal money for commercial airship construction, Hunsaker made a plea for just that. The "cheapest and most effective way" to ensure continued airship development, he said, was to pass subsidy legislation similar to the stillborn Merchants' Airship Bill. Yet now even an optimist like Hunsaker had to concede that bad weather in the Atlantic made safe commercial operations unlikely, and he shifted his attention to the Pacific, suggesting that a second hangar be built at the navy's West Coast base at Sunnyvale and that an airship be constructed specifically for the route to Hawaii. The hangar and airship would require a minimum government investment and provide additional facilities that could be used by the navy.[49]

Pressed to give the committee a cause for the accident, Hunsaker conceded that inadequate training and the rotation of officers from airships to sea duty may have been the reason for the *Akron*'s loss. He thought that "the continuity of [airship] commanders is desirable" to develop the experience and skills needed for the safe and efficient operation of airships. Similar to the basic seamanship that the masters of sailing ships learned in the nineteenth century, the experience gave today's ship commanders a feel for their vessels' limitations in adverse conditions. Hunsaker's conclusions reflected a nearly decade-long dispute between the Bureau of Aeronautics and the Bureau of Navigation, which was responsible for all naval personnel, over the assignment of aviation officers to sea duty. Moffett had argued that sending his officers to sea broke the continuity of their specialized aviation training, while successive chiefs of the Bureau of Navigation insisted that those who flew were naval officers first and that periodic sea duty was crucial to their professional development. Hunsaker was no more persuasive than his former boss was in resolving the issue, which bedeviled naval aviation throughout the interwar years.[50]

Hunsaker and other airship believers breathed a sigh of relief when the King committee's final report came out on 14 June. In brief, the committee found that the loss of the *Shenandoah* and the *Akron* were tragic reminders of the rocks and shoals of technological progress and that there was nothing fundamentally wrong with the navy's airship program. Following Hunsaker's suggestions, the committee recommended keeping Lakehurst open, placing the *Los Angeles* back in service as a training airship, and building a small airship to replace it. Most important, however, was the committee's belief that the large rigid airship still had a mission with the fleet and its recommendation that a new scouting airship be built as a replacement for the *Akron*. The King committee's findings were no reason for rejoicing in airship circles, because "the committee was not Congress," and its report was not nearly enough to overcome public and official skepticism about the cost and utility of the rigid airship. The *Macon* joined the fleet and Lakehurst stayed in business, but no money was forthcoming for new airships, and there was no hope for the resuscitation of the Merchants' Airship Bill.[51]

There is a sense of unreality about the development of the commercial airship in the United States, especially after the catastrophe of the stock market crash in the fall of 1929 and the depression that followed. How could someone as bright and knowledgeable as Jerome Hunsaker be lured into what we now know was a costly technological and economic dead end? There is no simple answer. In part, the engineer's perspective of the world was circumscribed in ways that sometimes led to conservatism and resistance to change. Like other professionals, his perceptions became internalized, resistant to "alien" practices, and locked in to a path from which it was hard to deviate.[52] In another sense, Hunsaker was caught up in the widespread popular enthusiasm aviation elicited in the 1920s and 1930s. Manufacturing and operating airplanes and airships engendered excitement that transcended the realities of the marketplace. Moreover, since leaving the Bureau of Aeronautics in 1923 he had been divorced from routine, hands-on aeronautical engineering and had no sense of the technological "revolution" that soon would come to airplanes and their propulsion.

So, too, was there an "internal logic" to the airship. To such rational people as Hunsaker the airplane had obvious strengths and weaknesses, and given the airplane's all-too-evident deficiencies in range and payload in the late 1920s and early 1930s, logic was, if anything, on the side of the airship for transoceanic flights.[53] Before the technical revolution in aerodynamics, materials, structures, engines, and propellers transformed the airplane in the mid-1930s, the airship's superiority in range and payload capacity made it supremely attractive to those in the navy looking for alternatives to surface ships for reconnaissance. Potential commercial operators recognized the airship as the only aircraft with the capability of carrying tens of passengers and tons of cargo across intercontinental distances.

Improbably, Hunsaker stubbornly adhered to the internal logic of rigid airship long after its demonstrated economic and technical shortcomings dictated its abandonment in favor of the far more efficient long-range airplane. He admitted his mistake, costly as it was in treasure and human life, but to the end of his life he regretted his failure to make the rigid airship something it could never be—a safe and efficient intercontinental transporter of people and goods.[54] With the navy proving the technology as the rigid airship's "launch customer" and Congress willing to pour millions into airship lines through airmail pay, it was reasonable to expect success. The loss of the *Akron*—and two years later its sister ship the *Macon*—did not cause Hunsaker to abandon his vision of the airship as a herald of worldwide aerial commerce. More perversely, though, Hunsaker's "logic" metastasized into a willful opposition to innovation that sometimes blinded him to technological and organizational change. Nevertheless, the immediate situation in the summer of 1933 led him to believe, correctly, that there was nothing much left for him to accomplish at Goodyear-Zeppelin, and he reluctantly decided to resign and seek an asylum of sorts in the safer and more secure academic world of MIT.

8.   MIT

J erome Hunsaker came away from his seven-year foray into the world of
   business with mixed emotions. The Bell Labs years, although a success in
   establishing AT&T's presence in the airways communication field and lucrative for Hunsaker personally, did little to satisfy his intellect or his
interest in aeronautical engineering. The Goodyear-Zeppelin experience was
bittersweet at best. In five years with the company, he had engaged in technical, economic, and political battles aimed at realizing the virtually limitless
possibilities of international air travel, emerging in the process as an experienced heterogeneous engineer with important contacts in the business and political communities. On the other hand, the crash of the *Akron* was a personal
loss that raised serious doubts about the future of the rigid airship in both military and commercial service. With some sense of relief, then, he took up his
new duties in the familiar and agreeable academic surroundings of MIT.

The opportunity for returning to Cambridge came when Karl T. Compton,
MIT's president, along with Vannevar Bush, dean of engineering, agreed that the
institute's Department of Mechanical Engineering needed to be more active in
such fields as thermodynamics, hydrodynamics, and materials. The imminent
retirement of older members of the department, Compton and Bush concurred,
opened a window for the appointment of new faculty members. Despite his
long and close ties to the institute, Hunsaker was not their first choice. They
looked initially at more "traditional" mechanical engineers before approaching
him in the late winter of 1933 to see if he might be interested in leaving
Goodyear-Zeppelin to take a position at MIT.[1]

After a meeting with Bush in March and a phone call or two, an understanding was reached that Hunsaker would take over as head of the department with primary responsibility for instruction in aeronautical engineering while retaining his connections with Goodyear through a "consulting arrangement." Paul Litchfield was not averse to the idea, which he thought might expand an already cordial relationship between Goodyear and MIT. The Executive Committee of MIT approved a salary of $12,000, and a formal offer went to Hunsaker on 11 April, but Compton feared people might think Hunsaker was leaving the company as a result of the *Akron* accident and counseled delaying an announcement. Later in the month, Hunsaker met with Compton in Washington and told him he would accept the offer and not make any statements until after the *Akron* hearings. Hunsaker recalled later in life that Compton's "judgment was good for me" and that he was grateful for his persuasiveness in urging him to come back to MIT.[2]

Understandably hurt when he learned of the decision to bring in Hunsaker as his replacement as department head, Edward F. Miller immediately offered his resignation. Bush exercised his executive authority to get Miller to agree to stay on through the summer while Hunsaker wrapped up his other responsibilities. C. Fayette Taylor, who had taught the department's aeronautical engineering classes for years, was not pleased with his removal from the aeronautical engineering program, but Bush assuaged his feelings by promising him exclusive control over all of the department's power plant work. Bush reassured Hunsaker that both men, despite the ill feelings, would work with him to reorganize the department and redirect its activities.[3]

Hunsaker wanted to move as quickly as possible in reorienting the department toward more serious research, yet he was acutely aware of Miller's and Taylor's sensibilities and did not want to make changes that were too drastic or too soon. He did think one faculty member, whose research on the flight of birds was "sophomoric," needed "to find other employment," and he remarked to Bush that Miller's courses "sound like popular mechanics." To soften the blow, he thought Miller ought to stay in his old office and continue teaching steam engineering, about which he had much "practical experience." As Hunsaker grappled with what to do with the former department head, Miller's health suddenly went into decline; he died in June. His death, though tragic, simplified the situation and allowed Hunsaker more flexibility in department decision making than would have been possible otherwise. Rather than wait much longer, Compton announced Hunsaker's appointment on 17 August 1933.[4]

Nearly a year passed before the Hunsakers bought a house in Boston. In the depths of the Great Depression home sales—especially upscale Long Island houses—were slow. Pending a buyer and a good price for the Glen Head property, the Hunsakers lived at the Hotel Sheraton on Bay State Road not far from MIT. Finally in May 1934 they found a large, four-story red brick house at 10 Louisburg Square in the Beacon Hill section of the city, notable among other things for having been the residence of Louisa May Alcott. Louisburg Square was one of the most fashionable neighborhoods in Boston, a verdant oasis designed to look like some of the finest residential squares in London and administered by the homeowners under terms written by the original nineteenth-century developer of the tract. Bought from an estate, the house was just affordable to the Hunsakers at the then-formidable price of $45,000, but it had great long-term investment potential. Hunsaker liked Louisburg Square. It was within walking distance of the MIT campus; in later years, as a leader of the proprietors' committee, he was known as the "mayor of Louisburg Square." Alice preferred something out of the city, where she would have had more room for her garden, but she was willing to make this concession as her part of the marriage bargain. Like the Glen Head property, the house at 10 Louisburg Square needed many repairs and upgrades, which the Hunsakers carried out over the next few years.[5]

What the Hunsakers gave up in open space they gained as members of Boston's social elite. Hunsaker joined the exclusive Saturday Club and the Thursday Evening Club, where members got together for dinner and conversation; Alice was active in one of the local gardening clubs. Unlike her husband, Alice had firm religious convictions, although she could see little difference among the various Protestant sects. Through her the family joined the King's Chapel Unitarian church, in part because she admired the character and preaching skills of its minister.[6]

The children's educations also reflected the Hunsakers' social standing. Sally, the eldest, was a senior at Sarah Lawrence College when the Hunsakers moved to Boston in 1933. In late 1936 she became engaged to David Swope, the son of Gerard Swope, president of General Electric; the couple married in King's Chapel in Boston in February 1937. The twins, Jerry Jr. and Peter, after completing the sixth grade at the Green Vale School on Long Island had enrolled at the Pomfret School in the fall of 1929. They then went on to Phillips Exeter Academy and to Harvard, where Jerry received his bachelor's degree in 1938. Peter attended Harvard for two years and transferred to MIT in 1937. Little Alice was not so little anymore, advancing through the awkward years of adolescence in day schools for girls in Boston and then going to Vassar College in the fall of 1940.[7]

Changes happened back home in Saginaw, too, during the 1930s. Walter had retired from the newspaper business in the 1920s but remained active until

Ten Louisburg Square. The Hunsakers moved to Boston when Hunsaker became head of the Department of Mechanical Engineering at MIT in the fall of 1933. (Courtesy author)

his health began to fail a decade or so later. Never truly close to his only son, Walter nevertheless enjoyed Hunsaker's visits to Saginaw and like most grandparents doted on the grandchildren. He suffered for years from atherosclerosis, probably exacerbated by a lifelong addiction to cigarettes, and ultimately had to have a leg amputated. His prospects of a quality lifestyle dimmed in 1939 when doctors determined that he might lose his other leg. By the late summer he had slipped into a coma, and doctors notified the family that he might not last more than a couple of days. Hunsaker left right away for Saginaw, arriving in time to see his father before he died on 28 September, a little more than a week past his eighty-second birthday. Hunsaker wrote to Gerard Swope that "my father's death was for him a welcome relief from a long and hopeless struggle and I think a relief for my mother also." He worried, though, about his mother now having to live alone in Saginaw and made a point of seeing her as often as his busy schedule permitted.[8]

Hunsaker's move to MIT came in the midst of strife in the nation and in American aviation. Not only had the country reached the depths of the Great Depression, but there had been the *Akron* disaster and the subsequent navy and congressional investigations in the spring of 1933. To make matters worse, a committee headed by Hugo L. Black, the Democratic senator from Alabama, began widely publicized hearings that fall, uncovering evidence that former postmaster general Walter Folger Brown and executives of the major companies holding airmail contracts had met in 1930 to divvy up the potentially lucrative transcontinental airmail routes. Condemned by some as "spoils conferences," the meetings appeared to be only the tip of the iceberg of widespread corruption and collusion between the Republican administration and big business. The Black committee recommended, among other things, that the government cancel all airmail contracts pending a thorough review of policies and procedures.

President Roosevelt agreed with the committee's findings and announced in February 1934 that the army air corps would henceforth fly the mails until new contracts could be drawn up and let under a fairer set of rules. The army's spirit may have been willing, but its equipment and training were not up to the task, especially in the face of some of the worst late-winter weather in decades. After a rash of accidents that took the lives of a dozen aviators, FDR ordered a new round of bidding on the airmail routes as an interim measure while Congress drew up revised legislation. Rather than solving the industry's woes, the Air Mail Act, coauthored by Black and passed in June, compounded them by locking the airlines into long-term contracts at rates far below what many thought was the minimum needed for economic good health. Stung by the whole episode, which gave the impression that the New Deal's aviation policy was in disarray, Roosevelt announced that in compliance with the provisions of the Air Mail Act of 1934 he was creating a special committee with broad powers to examine all of the problems facing American aviation.[9]

Headed by Clark Howell, the publisher of the *Atlanta Constitution,* the Federal Aviation Commission had wide-ranging responsibility for examining all aspects of aviation and making recommendations about policy for the next five to ten years. Hunsaker learned of his appointment to the commission on 1 July; other members were Hunsaker's friend Edward P. Warner, who served as the body's vice chairman, Franklin K. Lane Jr., a World War I aviator, and Albert J. Berres, who had been an official with the American Federation of Labor. J. Carroll Cone, assistant director of the Aeronautics Branch of the Department of Commerce, served as the commission's executive secretary. The response to Hunsaker's appointment varied. William B. Stout, noted for his metal aircraft designs, wrote to congratulate Hunsaker, stating that the Roosevelt administration had badly mishandled aviation policies and that he hoped Hunsaker would be able to "undo the damage" already done to the aeronautics industry. Samuel P. Wetherill, president of the Air Defense League, wanted a more independent body and criticized Hunsaker for being too close to Goodyear and the airship interests. Sensitive to this issue, Hunsaker informed Paul Litchfield that for the duration of his duty with the commission he would not do any consulting work for Goodyear-Zeppelin.[10]

In Europe until 26 July on business for Goodyear-Zeppelin, Hunsaker missed the Howell commission's preliminary meetings, which had begun on the tenth. Then, only a day after he arrived back in country, he and three other members of the commission—Warner, Lane, and Berres—took off on a whirlwind tour of aviation facilities in the United States and Latin America. Howell, meanwhile, headed for Europe to assess aeronautics policies and facilities in Britain, France, and Germany. Hunsaker's party left from Bolling Field in Washington in a Department of Commerce Ford Tri-Motor, with Langley Field as the first stop on their itinerary. After a tour of the NACA Langley Aeronautical Laboratory, the group flew on to Atlanta, Montgomery, Alabama, and Pensacola, where a navy transport picked them up and took them to Miami. From Miami on 31 July they flew to Brownsville, Texas, and then via Pan Am continued on to Cuba, Puerto Rico, Nicaragua, Honduras, Guatemala, and Mexico before landing at Randolph Field in Texas on 7 August. An American Airlines plane flew the group to the West Coast, where they visited airports, military installations, and manufacturing plants in San Diego, Los Angeles, and Seattle. On the return to the East Coast, there were stops at Kansas City, Chicago, Columbus, and Akron, with the exhausting, thirty-four-day, 13,000-mile (21,000-kilometer) odyssey finally winding up at Roosevelt Field on Long Island on 29 August.[11]

Nearly a month of planning preceded the opening of the commission's hearings on 24 September. Billy Mitchell, Hunsaker's old bête noir, was among the first of the major personalities to appear. Suffering from the heart disease that would kill him within eighteen months, Mitchell lashed out at the army and the findings of the recent board headed by Secretary of War Newton D. Baker. To his mind, the Baker board in recommending the establishment of a general headquarters air force and the consolidation of combat units under a single commander had done nothing to halt the decline of the army's air arm, which suffered from general neglect and from obsolete and dangerous equipment. The army's efforts to fly the mail were a "miserable mess," he declared, and only served to highlight the gross deficiencies of military aviation, adding that those who signed off on the Baker board report "ought to be kicked out of the service." The only solution, according to Mitchell, was a separate air force commanded by experienced flying officers. Much to Hunsaker's surprise and delight, Mitchell wanted more emphasis placed on big rigid airships, which he thought were needed not only to establish commercial air links to Europe and Latin America but also to rain tons of high explosives on Japan in the event of war in the Pacific.[12]

Charles A. Lindbergh left the unpleasantness of the New York extradition trial of Bruno Hauptmann, the accused kidnapper and murderer of his son, to appear before the Howell commission on 16 October. A large crowd packed into the hearing room in the Department of Commerce building to hear the great American hero condemn the administration's aviation policies. In response to Hunsaker's question about what "competitive conditions had been best for aviation," Lindbergh without hesitation replied that President Hoover and Postmaster General Brown had done the most to expand and promote civil aviation in recent years. Hunsaker next asked Lindbergh about the proper role of the federal government in regulating aviation. All passenger-carrying airlines had to meet minimum federally mandated safety requirements, Lindbergh noted, so a permanent aviation commission was necessary, but he maintained that the government should not intervene to the point of stifling competition in the industry. Finally, he predicted that the airlines would continue to require federal subsidies, at least for the immediate future.[13]

Starting in November, the commission spent a week in closed session dealing with issues concerning military and naval aviation and their implications for civil aeronautics, procurement, research, and technological change. A parade of officers representing the army's and navy's intelligence, logistics, training, engineering, and ordnance communities passed before the commission. Of major concern in light of Mitchell's testimony was the question of a unified, separate air force. To a man, naval officers opposed it, and most army people were at best lukewarm to the idea. Even the suggestion of centralizing aircraft procurement received little support. Congressman Carl Vinson outlined an ambitious naval building program that included two airships—one to replace the *Akron* and a smaller one to replace the *Los Angeles,* in addition to a big new aircraft carrier and a small flying-deck cruiser.[14] As the hearings ran their course, it became evident that those who expected the Howell commission to come up with a "silver bullet" to cure all the ills of commercial and military aviation were going to be disappointed.

The commission wound up its hearings during the third week of November, having heard nearly two hundred witnesses and recorded thousands of pages of testimony. There followed several weeks of hard work before the commission issued its final report on 22 January 1935. Befitting the president's mandate to look at all aspects of aeronautics, the commission's recommendations were broad ranging, while at the same time not diverging dramatically from the status quo. Possibly the most important element of the commission's report was the overwhelming support it gave to the continuation of federal regulation of commercial aviation. The airlines, like the railroads, were common carriers and as such were subject to the same regulation in the public interest; accordingly, the commission wanted the post office to relinquish its de facto regulatory role and the government to treat the airlines as it would other, more mature, forms of public transportation. In place of regulation through postal rates and route approvals, the commission recommended maintaining competition by the issuance of certificates of convenience and necessity, designating those carriers that met the government's criteria for service in the public interest.[15]

More controversial was the Howell commission's call for a permanent, independent agency responsible for overseeing civil aviation, including certification of airlines, rate setting, route approval, and the conditions of service for carrying passengers, mail, and cargo. The special circumstances under which civil aviation operated, the commission maintained, made such an agency necessary, although in the future the industry might be brought under the umbrella of a separate transportation body. The airlines welcomed the proposal, and there seemed to be considerable popular and congressional support for an independent board, but Roosevelt, hoping to curb the expansion of the federal bureaucracy and not wanting another agency outside the purview of the executive branch, opposed the idea. The concept never went away, however, finally reemerging as the Civil Aeronautics Authority (CAA) in 1938.[16]

As for military aviation, the commission determined to leave things as they were, much to the dismay of Mitchell and other air power advocates. The commission acknowledged that as military aircraft increased in effectiveness more independent operations were likely, possibly leading to fundamental reorganization of the army's air arm. For the time being, the commission recommended following through with the GHQ air force plan and other proposals put forth by the Baker board while calling for more interservice cooperation and continuing to study the "employment of air force as an independent striking unit." The commission also outlined the steps to be taken to integrate and standardize personnel, procurement, research, and training programs. Finally, the commission insisted that Mitchell's accusations about the inferiority of the design and construction of American military aircraft were, "if not actually false," then at least "misleading." Comparisons with European aircraft led to the conclusion that in all respects American combat aircraft were "at least the equal of any others in the world." Still, to maintain the country's position of technical superiority, much work and money would be needed over the next few years.[17]

Hunsaker's hand was most evident in the section of the Howell commission report dealing with airships. The losses of the *Shenandoah* and the *Akron,* the commission reasoned, were due to operational shortcomings and not to inherent problems in airship technology; in the future, the focus needed to be on improvements to navigation, weather forecasting, and training. The commission recommended proceeding with the navy's airship program and called for replacing the *Akron* as well as building a third airship to succeed the *Los Angeles.* As for the commercial airship, the commission determined that the volume of traffic on the North Atlantic route and the competition from fast European liners and the current German airship service dictated the government's investment. The Department of Commerce could administer a five-year $17 million experimental program using two airships on the transatlantic route. The airships and terminal would be built at government expense—possibly using Public Works Administration money—and then leased to a private operator. A successful demonstration of the service would encourage private investment and additional services across the Atlantic or the Pacific.[18] If nothing else, Hunsaker was persistent. In calling for a direct government investment in commercial airships rather than indirect subsidization through airmail contracts, the Howell commission's proposal was much more ambitious than that put forth in the Merchants' Airship Bill—and it stood even less chance of becoming reality.

Hunsaker's labors with the Federal Aviation Commission did not go unnoticed by the political leadership in Washington. His sweeping knowledge of aeronautics and reputation for integrity and hard work enhanced his position of respect and authority, which may have slipped during his years with Bell Labs and Goodyear-Zeppelin. His now quixotic advocacy of the rigid airship in the face of ever-growing odds against it masked a more evident reality. He may not have been aware of it from the security of his professorship at MIT, but by 1935 at age forty-nine his life and career had swung back toward public service, and he would be more in demand than ever as hard decisions had to be made about the government's role in aeronautical research and development.

For some time Hunsaker and others in the field had sought to create a professional organization to speak collectively for aviation, particularly those doing research in aeronautics. The American Society of Mechanical Engineers and the Society of Automotive Engineers had long welcomed aeronautics people into the fold but could not entirely satisfy their professional demands. Responding to the need, Hunsaker helped found the Institute of the Aeronautical Sciences and was elected as its first president at its inaugural meeting in New York in October 1932. It is easy to see Hunsaker's hand in the organization's principles, which emphasized that the "future development of aircraft depends on scientific research and engineering experiment applied to design" and the international exchange of information to promote "scientific progress" in aeronautics. He saw the institute casting a wide net, telling the press at the time that it "will bring in for the discussion of aeronautical problems, scientific experts in many fields." In January 1933, at the institute's Founder's Meeting at Columbia University, Hunsaker, Ludwig Prandtl, and Orville Wright became its first honorary fellows.[19]

In August 1933, not long after coming to MIT, Hunsaker took on the additional chore of editing the institute's new publication, the *Journal of the Aeronautical Sciences*. He told his friend Hugh L. Dryden at the Bureau of Standards, "I want to keep clear of descriptive articles from engineers who have built new airplanes or engines" and instead "put the emphasis on the ideas." For the most part he succeeded. In the first issue of the journal, which appeared in January 1934, he ran articles from Theodore von Kármán, Max Munk, Carl-Gustaf Rossby, and John Stack of the NACA Langley lab, among others, all written on a substantial—almost esoteric—theoretical level. Concerned that many of the highly theoretical pieces were beyond the comprehension of most workaday aeronautical engineers, he decided to run a greater percentage of more readable articles. In the fall of 1935, Caltech's Clark B. Millikan, the son of physicist Robert A. Millikan, wrote to Hunsaker to tell him that he was right in his assessment of the profession in America but reminded him of the importance of publishing scholarly articles, "if only for the purpose of giving the aeronautical engineers an idea of the activity which is taking place in this country along such theoretical lines." Hunsaker had planned to edit the journal for only a few issues while the institute found a permanent editor, only to find that, like many "temporary" jobs, this one lasted far longer than he had planned. He stayed on as editor until 1941.[20]

Hunsaker was also elected to the National Academy of Sciences (NAS) in 1935. One of the nation's most prestigious professional societies, the NAS had been founded in the nineteenth century specifically to advise the federal government on scientific and technical matters. By the 1930s its finances were in disorder. In 1940 the NAS appointed Hunsaker treasurer and member of the Finance Committee, with the specific task of reviewing the academy's investments. His personal expertise in finance and his connections with the Boston moneyed elite served him well in this role; in short order he reduced the academy's holdings in stagnant railroad stocks and sufficiently diversified its portfolio to bring significantly higher returns and financial stability.[21]

Despite Hunsaker's understanding with Compton that the MIT position would not interfere with his responsibilities outside the institute, departmental politics and the day-to-day decisions that had to be made about professors, students, and curriculum inevitably intruded. Fortunately, Hunsaker had help from Richard H. Smith, who functioned as a sort of executive officer in the department. The department's core faculty was dedicated to scholarship and teaching in aeronautical engineering. Among them were Otto C. Koppen, who had worked with Fairchild, Ford, and the army at McCook Field; Shatswell Ober, who did much of the laboratory work using the institute's wind tunnels; and Edward S. Taylor, an expert in the field of aircraft propulsion and power plant research. Largely due to Hunsaker's influence, C. G. Rossby left the Weather Bureau after working on the West Coast model airline project to join the MIT faculty in 1928, adding the latest meteorological theories to the school's curriculum.[22]

Hunsaker also played an active role in changing MIT's undergraduate and graduate curricula in aeronautical engineering. Over the first three years of his term as head of the department, the undergraduate curriculum that had been established in 1926–27 underwent revisions to bring it more in line with the standards of other engineering curricula at the institute. Emphasis was placed on fluid dynamics, thermodynamics, and electrical engineering, and room was made for more electives. The biggest change in the graduate program came in 1934 with the appointment of three graduate students as research assistants. The idea was to provide people to support continuing research projects as well as to reward outstanding students with money and work experience. At the graduate level, Koppen's class on airplane design reflected some of the new thinking that emphasized the airplane as part of a larger technological system. Hunsaker himself developed a new offering on fluid mechanics, which had application to aerodynamics, hydraulics, and thermodynamics, and taught it to juniors during the 1934–35 academic year.[23]

One of Hunsaker's priorities in reinvigorating MIT's aeronautical engineering courses was attracting new faces, a task eased by his many connections in business, government, and academe. In December 1933 he wrote to Clark Millikan that even though he had been at MIT only for a couple of months, "it is long enough to begin to see opportunities to make some improvements. The principal ones seem to lie in the addition to the staff of a few younger men in the fundamental fields of engineering to bring to Technology [MIT] a different background and viewpoint." He wanted Millikan to come to MIT but could not persuade him to leave sunny Pasadena. With other bright young prospects Hunsaker was more successful. Two additions to the faculty in 1935 were Charles Stark Draper, a research associate who specialized in aircraft instrumentation and later made a name for himself in inertial guidance systems for aircraft and missiles, and Heinrich Peters, a Göttingen student with a superb understanding of the latest wind tunnel designs.[24]

Peters was the key figure in the department's efforts to acquire a state-of-the-art wind tunnel. When Hunsaker returned to MIT, the department had four tunnels, all low-speed, open-circuit designs. Hunsaker's four-by-four-foot (1.2-by-1.2-meter) tunnel of 1914 had been augmented in 1921 by a higher-speed tunnel, but his old device continued to soldier on for undergraduate instruction. The four tunnels at MIT could not achieve speeds in excess of 150 miles (241 kilometers) per hour, were unsuitable for research on many contemporary aircraft, and were totally inadequate for the aircraft expected before the end of the 1930s, which would likely approach speeds of 400 miles (644 kilometers) per hour. Compton underscored the institute's deficiencies in 1936, insisting that unless MIT built a new tunnel the aeronautical engineering program would be relegated "to an inferior and less effective position."[25]

Meanwhile, plans for a new tunnel were taking shape. Closely adhering to a Göttingen design, Peters proposed a closed-circuit tunnel with a 2,000-horsepower electric motor driving a thirteen-foot (four-meter) variable-pitch propeller to provide speeds up to 400 miles (644 kilometers) per hour. One important feature of the tunnel was that it could be pressurized to four atmospheres or depressurized to one-quarter of atmospheric pressure to simulate high-altitude conditions. Additionally, the tunnel's seven-and-a-half-by-ten-foot (2.3-by-3-meter) elliptical test section permitted the use of larger test models than was usual while maintaining a reasonable total area. Much of the tunnel's detail design was handled in house, with John R. Markham designing the six-component balance and Charles Stark Draper providing the instrumentation. In February 1937 President Compton announced MIT's commitment to build the facility, to be named the Wright Brothers Memorial Wind Tunnel, followed in May by the MIT Corporation's approval of the project.[26]

Even before the tunnel received the go-ahead, Hunsaker had approached his many connections in the aeronautical community to raise more than $90,000 toward the estimated $230,000 total cost of the project. Among those who con-

The Wright Brothers Memorial Wind Tunnel under construction at MIT, 1938. Hun-saker was instrumental in raising the money for the tunnel, which became a premier aerodynamics research facility in the 1940s and 1950s. (AA 308, Courtesy MIT Museum)

tributed to the tunnel were Vincent L. Bendix, Edward A. Deeds, Godfrey L. Cabot, Frederick B. Rentschler (president of United Aircraft), George J. Mead (a United Aircraft vice president and MIT graduate), and the Curtiss-Wright Corporation. In October 1937 Jackson and Moreland, a local engineering company with close ties to MIT, won the general contract, with Pittsburgh–Des Moines Steel Company responsible for the fabrication of the steel shell of the tunnel and Westinghouse for the electrical equipment. Dedication of the new facility, conveniently located adjacent to the Guggenheim Building, took place in ceremonies on 12 September 1938 held in conjunction with the fifth annual meeting of the International Congress of Applied Mechanics.[27]

The Wright Brothers Tunnel was one of the best in the country at the time. The variable density feature made possible high Reynolds numbers, and the size of the test section allowed the use of relatively large models, yielding more accurate data. Although the tunnel could not attain supersonic speeds, it did have the capability of studying the problems of compressibility and shock waves using model propellers driven at high tip speeds. Not surprisingly given the strong industry support for the Wright Brothers Tunnel, much of the early experimentation centered on projects sponsored by various aircraft and engine manufacturers.[28]

Then as now, those in academic engineering positions spent much of their time outside the classroom doing consulting work for industry. Hunsaker was no exception, serving in the 1930s as a member of United Aircraft's Technical Advisory Committee. From 1935 to 1936 he participated in an intensive year-long study of transport aircraft, much of which dealt with the long-range flying boat. In the mid-1930s Pan Am was interested in the work of United's Sikorsky subsidiary on big flying boats capable of spanning the Atlantic nonstop. As Hunsaker knew from his analyses of airship operations, Sikorsky and other companies had their hands full designing airplanes with the performance needed for commercially profitable transatlantic flights, where they would have to take off with full loads under adverse sea conditions.

Hunsaker understood, as did all aeronautical engineers, how crucial weights and wing loadings were to aircraft performance. Sikorsky's 38,000-pound (17,237-kilogram), four-engine S-42 flew for the first time in 1934. Its wing loading of 28.5 pounds per square foot (139 kilograms per square meter) was "mind-boggling in an airliner of this date." The S-42A was heavier and had a wing loading of about 30 pounds per square foot (146 kilograms per square meter), but Hunsaker warned that newer designs would need even higher wing loadings to ensure commercial feasibility, probably on the order of 32 or more pounds per square foot (156 kilograms per square meter). Some performance gains were also possible using flaps and other high-lift devices and by fitting more powerful engines and more versatile propellers, but higher wing loadings were the real keys to success.[29]

During the 1930s American aeronautical engineers achieved major advances in aircraft performance; these developments were manifested in the big flying boats built by Sikorsky, Martin, and Boeing. It has been emphasized that these aircraft represented a revolution in design and construction overshadowing the famous Boeing 247 and Douglas DC-1,2,3 series of airliners. In contrast to the modest performance of the twin-engine Boeing and Douglas airplanes, which were limited to no more than twenty-one passengers and ranges of about 400 miles (644 kilometers), the four-engine flying boats could carry up to thirty passengers over distances of more than two thousand miles (3,219 kilometers). The secret to this remarkable performance was the incorporation of new and stronger aluminum alloys, close attention to structural details, and most important, wing loadings in excess of thirty pounds per square foot (146 kilograms per square meter). One result was that the United States stole a march on the British, establishing regular transatlantic passenger and mail service in the late 1930s.[30]

Some of the results of Hunsaker's United Aircraft study appeared in an article in the December 1936 issue of *Technology Review,* coauthored with George Mead. Avoiding sweeping generalizations or "pie-in-the-sky" dreams, the article emphasized the importance of economic considerations in aircraft design. Hunsaker and Mead's "Tomorrow's Airplane" was a twenty-passenger, low-wing, all-metal, four-engine monoplane with a maximum takeoff weight of 40,000 pounds (18,144 kilograms), a cruising speed of 225 miles (362 kilometers) per hour, and a range of 1,250 miles (2,012 kilometers). A blend of old and new, the airplane had high-aspect-ratio wings with a loading of more than thirty pounds per square foot (146 kilograms per square meter). Lacking a pressurized cabin, it could not routinely fly above the turbulent weather of the lower atmosphere, and its landing gear, though fully retracting, featured a conventional tail wheel.[31]

It is instructive to compare Tomorrow's Airplane to the Douglas DC-4E and the Boeing 300, both of which were in the design stages when the Hunsaker and Mead article came out. The Douglas flew first, in June 1937. Like Tomorrow's Airplane, the DC-4E was a low-wing monoplane with four engines; it had a similar cruising speed, but the Douglas was a much larger aircraft, with a maximum takeoff weight 50 percent more than Tomorrow's Airplane, accommodating forty-two passengers (more than twice as many as the Hunsaker-Mead concept) and having a tricycle undercarriage. The DC-4E's wing loading of 28.5 pounds per square foot (139 kilograms per square meter) was only marginally less than Tomorrow's Airplane. With 1,450-horsepower Pratt and Whitney R-2180 engines, the Douglas was superior to Tomorrow's Airplane, which had 1,100-horsepower engines. The most dramatic difference between the Hunsaker-Mead ideal airplane and Douglas's real airplane was range: The DC-4E could fly 2,200 miles (3,540 kilometers), nearly 1,000 miles (1,600 kilometers) more than Tomorrow's Airplane. United Airlines took delivery of the one and only DC-4E and evaluated it for service in 1939, only to have the airlines declare that it was "too much airplane too soon."[32]

In some ways Boeing's Model 300, which flew for the first time in December 1938 and was later developed into the Model 307 Stratoliner, was closer to the Hunsaker-Mead airplane. Its 42,000-pound (19,051-kilogram) maximum takeoff weight was similar, as was its cruising speed (220 miles [354 kilometers] per hour), range (1,750 miles [2,816 kilometers]), and engines (four 1,100-horsepower Wrights). Its wing loading, at a little more than twenty-eight pounds per square foot (138 kilograms per square meter), was about the same as that of the DC-4E. On the other hand, the Boeing could carry thirty-three passengers, and most important, it featured a pressurized cabin that made possible high-altitude operations. Delivered in small numbers to Pan Am and Transcontinental and Western Air, the Model 307 Stratoliner saw only limited service before being taken up as a military transport during the war.[33]

If Tomorrow's Airplane stayed close to the "state of the art" in the mid-thirties, the "Future Airplane" was more ambitious. Hunsaker and Mead suggested that this aircraft might be smaller and lighter than Tomorrow's Airplane, incorporating advances in wing design to reduce skin friction and promote laminar flow. Wing loadings would also go up, although not dramatically more than on Tomorrow's Airplane. In the absence of any breakthroughs in aerostructures, Hunsaker and Mead concentrated on improvements to power plants, contending that the engines of the Future Airplane would achieve greater power and fuel efficiency through the use of improved cooling and supercharging. The design also incorporated a pressurized cabin, allowing the airplane to operate above 25,000 feet (7,620 meters), where it could take advantage of high-velocity stratospheric winds and avoid lower-altitude storms and turbulence, particularly on longer-distance flights. As with Tomorrow's Airplane, economic factors weighed more heavily than cutting-edge technology; Hunsaker and Mead wanted an airplane that would reduce operating costs, enhance passenger comfort, and give the airlines something with which they had a reasonable chance of making money — all realistic considerations in the mid-1930s.[34]

Despite the commercial advantages of long-range land planes and flying boats, Hunsaker was still not prepared to give up on the airship for transoceanic travel. The navy made Lakehurst available as a terminal for the LZ.129, now named the *Hindenburg,* which began regular transatlantic service in 1936,

raising Hunsaker's and others' hopes that one day the United States would
be a more active participant in commercial airship operations. In September
1936 Hugo Eckener announced that Luftschiffbau Zeppelin planned to build
two more airships and met with President Roosevelt, who thought the German
flights might revive American commercial interests. The next month, Hunsaker
used his contacts with the Zeppelin organization to help arrange a ten-hour
publicity flight by the *Hindenburg* from New York to Boston and back. Enthusiasm for the airship was renewed by changes in both its name and its leadership:
the International Zeppelin Transport Company became the American Zeppelin
Transport Company, with Edward Deeds as chairman, Edward Farley as president, and Eckener's close associate F. Willy von Meister as a vice president.
Hunsaker and Litchfield were two of the firm's many directors.[35]

Lured by the taste of hearty lager and Oktoberfest hospitality as well as the
opportunity to see Germany's aviation manufacturing facilities and laboratories, Hunsaker jumped at the chance in 1937 to deliver a paper on transatlantic
air transport at an international conference sponsored by the Lilienthal Gesellschaft in Munich. He sailed from New York early on 29 September aboard the
Hamburg–America Line's *Europa,* arriving in Bremen on 5 October after a
seven-day passage. Escorted by his hosts, he made his way to Munich for the
opening of the conference at the Deutsches Museum on the eleventh.[36]

Hunsaker's paper synthesized the work he had done on commercial airship
operations for Goodyear-Zeppelin and his studies of transport airplanes for
United Aircraft. Despite the loss of the *Hindenburg* in a fiery crash at Lakehurst
in May, Hunsaker remained convinced that an airship of about 9.2 million cubic
feet (260,515 cubic meters; 28 percent larger than the *Hindenburg)* and inflated
with helium could be safe and economical in regular transatlantic service. In
contrast, he estimated that a 100,000-pound (45,359-kilogram) flying boat
would need to haul at least fifteen passengers and a ton (908 kilograms) of
cargo to be profitable in nonstop service. A land plane, using some sort of assisted takeoff or in-flight refueling, could carry twenty-five passengers across
the Atlantic. Regardless, wing loadings had to be increased, possibly as high
as 50 pounds per square foot (244 kilograms per square meter). No airplanes
capable of profitable nonstop transatlantic flights then existed, but Sikorsky's
most recent flying boat, the S-42B, with a wing loading of 33.5 pounds per
square foot (164 kilograms per square meter), came close, and American manufacturers were working on airplanes with loadings approaching the 50-pound-
per-square-foot (244-kilogram-per-square-meter) ideal. Hunsaker tried to make
the point that the airship and the airplane were complementary—the former
providing two-day service in comfort rivaling that of the best ocean liners and
the latter offering less comfort but twice as much speed.[37]

Hunsaker at the meeting of the Lilienthal Gesellschaft, Munich, 1937, at which he delivered a paper on transoceanic air transportation. To his right is Frank Caldwell of United Aircraft's subsidiary Hamilton Standard Propeller Company. Caldwell was the engineer and designer of an efficient variable-pitch propeller. (Courtesy JCHP/MIT)

During his ten days in Germany, Hunsaker visited the aerodynamic research facilities at the University of Göttingen, the Luftschiffbau works at Friedrichshafen, and the Heinkel aircraft manufacturing facility at Warnemunde before going on to Paris and London. During a week's stay in Britain he saw familiar faces at Oxford, Cambridge, and the University of London, toured aircraft factories, and inspected the most modern wind tunnels at the National Physical Laboratory at Teddington and the Royal Aircraft Establishment (RAE) at Farnborough. He left on the *Europa* on 27 October and returned to the United States on 1 November, deeply impressed by the scale and the quality of aeronautical research being done throughout Europe and in Germany in particular.[38]

More studies of flying boat designs for United Aircraft finally led Hunsaker reluctantly to give up on the airship and convinced him of the commercial viability of the airplane in transoceanic service. When Mead broached the idea of Sikorsky entering a twin-engine design into a navy competition for a new generation of flying boats in 1937, Hunsaker conceded that there were advantages of lower cost and reduced maintenance but that the navy was likely to insist on four engines for greater performance and reliability. It made no sense to Hunsaker for Sikorsky to concede what it did best simply because its competitors were offering two-engine alternatives to the company's proven four-engine design. It was better for the company to stay with "fully developed" engine-propeller combinations and improve the aerodynamics of its existing XPBS-1 flying boat design. Sikorsky could do much to reduce drag by "cleaning up" the flying boat's wings and fuselage, using flush rivets and retractable wing floats, and it could increase range and payload with high-lift devices and increased wing loading. Whatever Sikorsky decided to do, Hunsaker concluded that the company should not wait and see what Consolidated and Martin came up with before moving ahead with its own proposal.[39]

At Pan Am's request, Sikorsky completed a series of preliminary design studies for a gigantic transatlantic flying boat in January 1938. Mead met with Hunsaker in Cambridge to brief him on the results and invited his comments. Hunsaker found it intriguing that Sikorsky had not limited itself to flying boats and had suggested that a high-speed land plane might meet Pan Am's needs better. Sikorsky's studies meshed with some of the results of Hunsaker and Otto Koppen's findings that flying boat designs could never overcome inherent drag penalties due to the shape and size of their hulls. Nor did Hunsaker like Sikorsky's plans for using six engines; four, he thought, were sufficient even for the largest aircraft then under consideration. Finally, Hunsaker questioned whether Pan Am really needed a huge flying boat. Would it not be better to provide two airplanes of about 80,000 to 100,000 pounds (36,287 to 45,359 kilograms) maximum takeoff weight, which could meet the airline's requirement for carrying a hundred passengers across the Atlantic, while providing more convenience and flexibility?[40]

By 1938 Hunsaker at last realized that the airship's time had come and gone. For whatever reason—professional "lock-in" or the internal logic of the airship—he had failed to comprehend that people valued speed above all in air travel and that the airplane could now offer safe, fast, and relatively inexpensive intercontinental transportation. Now it was clear that the flying boat and perhaps the long-range land plane had rendered the airship obsolete.

Although some mourned the passing of those noble but fragile craft, Hunsaker never looked back, except to lament in later years that he had expended so much time and effort on a dead-end technology. On a personal level, he said that he felt "a sense of responsibility for those men going into [airships], because they thought I knew what was what, and they took my word for it." He also regretted that the expected payoff of the airship as a commercial entity never materialized: "It was illusory. I thought, and many others did, along with [Paul Litchfield], that we could establish a transatlantic air passenger service," but the airship, despite all its promise, never met its proponents' expectations. As "good engineers," he and others should have realized by the mid-1930s that the airplane—especially the four-engine flying boat—could now carry a useful payload over transoceanic distances and that by the time the airship became operational in passenger service it had lost its initial advantages. For Hunsaker, it was a painful realization: "We were wrong, economically and commercially."[41]

Hunsaker's thinking about transatlantic air operations underwent further changes as a result of a study he undertook for American Export Airlines. In late December 1939 Hunsaker met with John E. Slater, the company's vice president, who was looking for technical assistance and exploring ways to break Pan Am's monopoly on transoceanic air travel. Earlier in the year, Pan Am had begun regular service across the Atlantic with the Boeing 314, an 82,500-pound (37,421-kilogram) flying boat. Slater wanted to counter Pan Am with nonstop flights from either Boston or New York to Lisbon, with connections through the Mediterranean and Black seas, using something comparable to Pan Am's Boeing or the Martin M-130 four-engine flying boats. The airplanes did not have to carry large payloads, because the company was largely interested in establishing a presence on the routes and giving passengers a potential alternative to Pan Am. Slater needed information on various types of airplanes and whether or not they had the performance to secure Civil Aeronautics Authority certification when his company applied for the routes later in the year.[42]

This was familiar territory for Hunsaker, who through his Goodyear-Zeppelin and United Aircraft studies knew as much as anyone about the feasibility of transoceanic air services. He understood that the big flying boats generally had the required performance but that they were all deficient in some regard. For the American Export study, Hunsaker gathered data on the Martin 162 twin-engine flying boat (later to become the highly successful PBM Mariner), the four-engine Martin M-130 flying boat, the British Albatross four-engine land plane, and the Boeing 314, Consolidated PB2Y-2, and Sikorsky PBS-1 designs. On the West Coast he visited the Boeing and Lockheed plants, coming away dismayed by the size and expense of the Boeing 314 and favorably impressed by a sleek, Lockheed four-engine L-44 Excalibur land plane that the company was working on. His analyses showed that Boeing was too optimistic about the range of the 314 and that, similarly, Martin had oversold the navy on the endurance of the 162. Although the twin-engine Martin looked like "the best available," it was still, he said, "short of what you really need," largely due to the immaturity of its Wright R-2600 Cyclone engines. Moreover, Hunsaker doubted that the Civil Aeronautics Authority would allow passengers to fly across the Atlantic in a twin-engine aircraft.[43]

The deeper into the analysis he got, the more Hunsaker became certain that a four-engine land plane best suited the airline's needs. He urged Slater to con-

tact Robert Gross at Lockheed to have a look at the L-44, which would probably fly sometime in 1940. With modifications, the airplane could achieve a
3,600-mile (5,794-kilometer) range, allowing nonstop operations across the
Atlantic, albeit with only a small payload. He also thought the Boeing 307 Stratoliner had possibilities, although he wondered that it might be too large for
this application. In a letter to Claire L. Egtvedt at Boeing, Hunsaker wrote that
he thought the 314 was too large and costly for what his client wanted but
that the 307, slimmed down to about 50,000 pounds (22,680 kilograms), might
be something the company would be interested in.[44]

Hunsaker's thoughts were consistent with his *Technology Review* ideas about
how less was more in transport aircraft. His advocacy of land planes may seem
to run contrary to those who insisted on the big flying boat for long over-water
flights, but the thinking was actually strikingly similar. Like the designers of
the Sikorsky, Martin, and Boeing flying boats, Hunsaker was keenly aware of the
importance of weight control and the advantages of high wing loadings. Even
though high wing loadings meant longer runways, concrete, Hunsaker knew,
was less costly than the aerodynamic and hydrodynamic shortcomings inherent
in the large flying boat.

But Slater remained unconvinced that an advanced land plane was his company's airliner of choice, and Hunsaker had to give in. As a consultant with
United Aircraft, which owned Sikorsky, he was familiar with the PBS-1, which
despite range and payload deficiencies was in his mind the "best flying boat
produced in recent years." Based on Hunsaker's recommendation, Slater decided to order three Vought-Sikorsky four-engine VS-44 flying boats from
United Aircraft. The commercial version of the PBS-1, the VS-44 had a range
of 3,800 miles (6,115 kilometers) and could accommodate up to forty-two passengers on shorter flights. After many delays, the company finally won approval to fly across the Atlantic, beginning nonstop service from New York to
Ireland with the VS-44 boats in June 1942.[45]

Hunsaker's persistent lobbying for a land plane culminated in 1941 with an
intensive comparative study of the land plane and the flying boat for American
Export. The company asked Hunsaker for the study in March 1941 to resolve
conflicting information on the two aircraft types; Sikorsky insisted that the co-
efficient of drag in flying boats and their percentage of structural weight versus
gross weight decreased as they became larger, eventually rivaling or exceeding
that of any land plane. Others argued that the best the flying boat could do was
to equal the performance of the land plane, which offered the added advantage
of being able to operate year-round, of having lower maintenance and support
costs, and of flying directly to inland terminals. An American Export executive
acknowledged that "the psychological effect on passengers of flying over vast

stretches of water in a land plane cannot be overlooked," but he understood that it was unlikely that a multiengine land plane would ever have to make a forced landing at sea and that even for a flying boat such an event was still dangerous, especially in heavy weather.[46]

Because of his growing workload and other commitments outside MIT, Hunsaker had to turn over much of the work to his colleague Otto Koppen, but the final product generally reflected his views that the flying boat was, if not obsolete, then certainly obsolescent. The study assumed that the aircraft would be large—about 100,000 pounds (45,359 kilograms) maximum takeoff weight— and would have four engines, a range of 5,000 miles (8,047 kilometers) in still air, and a wing loading of at least forty-five pounds per square foot (220 kilograms per square meter). The flying boat could be built about 5,000 pounds (2,268 kilograms) lighter than the land plane, due to the absence of landing gear, but it was not as "clean" aerodynamically, and its weight advantage disappeared at cruising speeds above 150 miles (241 kilometers) per hour. With a payload of 10,000 pounds (4,536 kilograms), the land plane could achieve a cruising speed of 236 miles (380 kilometers) per hour, while the flying boat flew at only 208 miles (335 kilometers). As for relative safety, the report concluded that with modern engines there was little danger of a forced landing with either type. In later years, American Export abandoned the flying boat and turned to four-engine land planes, acquiring four-engine Douglas C-54s for transoceanic flights.[47]

Through his work on United's Technical Advisory Committee, Hunsaker studied recent developments in propeller technology by Frank Caldwell at United's subsidiary Hamilton Standard Propeller Company in East Hartford, Connecticut. Promising major gains in aircraft performance because it allowed the engine to run at optimum speeds at various altitudes and temperatures, the variable-pitch propeller was a tough technical challenge. Caldwell's design had a simple and rugged hydraulic actuating system, but the electric mechanism developed by Curtiss-Wright was an attractive alternative.[48]

At George Mead's request Hunsaker provided an in-depth appraisal of Hamilton Standard's propeller research program. In a long memorandum on 7 July 1937, Hunsaker stressed that Hamilton Standard had achieved a position of leadership in the field by forward-looking research, positioning the company to exploit the market once airplanes attained speeds high enough to warrant the use of variable-pitch propellers. Absolutely vital was "research in advance of demand," whereby the company looked beyond immediate needs and profits to stay comfortably ahead of its competitors.[49]

Finding that "in spite of all the theories and the flight and wind tunnel test data," engineers still applied a disconcerting amount of guesswork to propeller design and construction, Hunsaker offered specific suggestions about what Hamilton Standard could do in its research program. Initially, he thought there were ways to improve the existing Caldwell design by using gears to achieve a greater range of pitch and to simplify the propeller's self-feathering capability. Four-blade and contrarotating propellers offered possibilities, as did materials other than forged duralumin for propeller construction. The company also had to address the problem of compressibility at high propeller tip speeds. An "elaborate" series of wind tunnel tests with full-scale engine-propeller combinations would allow the company to improve hub designs, investigate new materials, and optimize blade shapes and numbers of blades. Hunsaker concluded that either MIT or the NACA Langley lab might be the best place to help the company with a rational, long-term propeller experimental program.[50]

Caldwell's response to Hunsaker's memorandum reflected his experience with propellers and their operation. He and Hunsaker were in "fairly good agreement" on propeller research, but he thought some of Hunsaker's points needed clarification. The hydraulic system used in Hamilton Standard variable-pitch propellers, Caldwell agreed, might not provide a full range of control, but it had proved robust in service. Caldwell was most concerned about vibration, which Hunsaker had all but ignored in his report. Although four-blade, and even six-blade, propellers were a good idea, contrarotating propellers had inherent vibration problems that could not be easily overcome. Caldwell concurred with Hunsaker that new materials might bring significant improvements, but he cautioned that tests of welded steel blades had resulted in severe vibrations even at moderate speeds. Finally, Caldwell agreed with Hunsaker about the need for more research into compressibility at high tip speeds, but he added that even an efficient high-speed blade design might still encounter vibration problems.[51]

United Aircraft had to reevaluate its research program when Hunsaker informed the company that Curtiss-Wright was modifying one of its existing tunnels for advanced propeller experiments. The specter of one of United Aircraft's principal competitors gaining an advantage in the propeller field was sufficient to push Hamilton Standard toward building its own tunnel. Hunsaker wrote to Caldwell offering help on some of its specific features. A 200-mile (322-kilometer) -per-hour, closed-circuit tunnel with an eighteen-by-eighteen-foot (5.5-by-5.5-meter) test section would be suitable for most propellers then in existence or planned. The tunnel should be of a closed-throat design, which restricted the size of the engine-propeller combination and limited access to the test section but precluded having to maintain a constant pressure within the test house, thus saving considerably on construction and operating expenses.[52]

Hunsaker at his desk in the Guggenheim Building, MIT, 1940. (JCH 9, Courtesy MIT Museum)

With Hunsaker's approval, Jackson and Moreland completed a study of the proposed tunnel in February 1941, finding that it would help the company determine the best engine-propeller installations early in the aircraft's design phase, explore unconventional designs, improve power plants in existing aircraft, and verify engine-propeller data from other sources. Jackson and Moreland proposed a closed-throat, 210-mile (338-kilometer) –per-hour tunnel with an eighteen-by-eighteen-foot (5.5-by-5.5-meter) test section. The proposed tunnel sharply differed from Hunsaker's in being an open-circuit type, Jackson and Moreland reasoning that because a power plant tunnel typically needed about a quarter of its air renewed each cycle, it made more sense—and it was less expensive—to make it an open-circuit design. Hunsaker liked the proposal in general but changed his mind on the closed-throat test section, deciding that an open-throat design was better for moving big equipment in and out of the test section, considerably reducing labor costs. United Aircraft decided to stay with the closed-throat option and in most other respects adhered to the Jackson and Moreland plan. To avoid conflicts of interest, Hunsaker severed his formal connections with United Aircraft at the end of June 1941, when he returned to active duty in Washington as a reserve captain to reorganize the navy's research and development program.[53]

Hunsaker had found a welcome sanctuary at MIT when his career at Goodyear-Zeppelin had reached a dead end with the abortive commercial airship project. His headship of the institute's mechanical engineering department, and after 1939 the new Department of Aeronautical Engineering, brought him back to his professional roots and put him in touch with the most promising aeronautical developments. United Aircraft and other consultancies were valuable adjuncts to his academic duties and indicators of his maturation as a heterogeneous engineer. While cultivating already close ties with the aircraft industry, Hunsaker used his advisory work to stay involved in the latest aircraft and engine development programs, to steer lucrative research projects to Cambridge, and to reappraise the relative merits of land-based airplanes, flying boats, and airships in long-distance commercial operations.

Industry connections cut both ways. When Hunsaker accepted responsibility for much of the nation's aeronautical research and development programs during World War II he knew who to approach in aircraft and engine manufacturing to get things done. At the same time, his close association with the private sector made it hard for him to follow his instincts in adhering to a wider public interest. Unavoidably, Hunsaker found himself weighing the immediate needs of industry for development and testing against the long-range basic research he knew was essential for the nation's technological well-being.

9.　　　　War and the NACA

A succession of crises swept the world during the 1930s, beginning with the Japanese invasion and occupation of Manchuria in 1931, continuing through the middle of the decade with the Italian conquest of Ethiopia and the Nazi remilitarization of the Rhineland, and climaxing in 1936 with the tragedy of the Spanish Civil War. Another devastating global conflict seemed an all-too-real possibility.

Like many scientists and technologists in the United States, Jerome Hunsaker was aware that security in great part depended on rational research and development policies, especially in aeronautics, where Germany seemed to have gained an advantage over the Western democracies. He and others worried about a rising Germany armed with superior aircraft emanating from an aggressive, state-funded aeronautical research program. The Czechoslovakian crisis of 1938 and the "appeasement" of Hitler at Munich by Britain and France, which feared the potential might of the German air force, did nothing to allay Hunsaker's fears. At home, President Roosevelt responded to the alarming events in Europe with the proclamation of a limited national emergency and called for strengthening America's armed forces.

Hunsaker shared his concerns about defense with fellow engineers and students on a nationwide speaking tour for Sigma Chi, the Scientific Research Society, in the spring of 1939. On 6 May, he spoke at Alabama Polytechnic Institute (now Auburn University), his last stop before returning to MIT. He told his audience that although the United States had a clear lead in commercial airplanes, "there is reason to believe that the latest types of fighting and bombing airplanes produced in Europe fly faster and higher" than anything in the American inventory. The expansion of the German Luftwaffe had "dramatically readjusted the balance of power," with potentially serious consequences for the United States. Hunsaker concluded that the country "must vigorously pursue the application of aviation to national defense" and accelerate its research efforts to remain "in the forefront of aviation development."[1]

Most important, as a member of the NACA and then as its chairman, Hunsaker saw an opportunity to marshal his and others' technical expertise in reversing the decline of basic aeronautical research in the United States. Hunsaker's experience with the NACA illustrates the issues relating to the role of scientists and engineers in establishing and implementing national security policy within the military-industrial-university complex and provides insight into the political, economic, and military forces that drove technological change during the war years and after.

Hunsaker's association with the NACA began when his MIT study of the longitudinal stability of airplanes formed part of the first technical report published by the NACA in December 1915. He had also been one of the navy's members on the NACA Main Committee in 1922–23, before resigning to take the attaché job in London. Back from England in 1926, he told George Lewis, the NACA's director of research, that "if you have a vacancy on the Committee in [the] future—consider me available." The opportunity did not come up until August 1938, when Hunsaker returned to the NACA as a member of the Committee on Aerodynamics, chaired by his friend Edward P. Warner.[2]

By the late 1930s, the NACA had earned a well-deserved reputation as the nation's premier organization devoted to aeronautics research. An independent entity within the federal bureaucracy, the NACA differed from other agencies in stressing collegiality over hierarchy and governance by consensus. At the top was the fifteen-member Main Committee, which set broad policy guidelines, selected areas of research, and acted as an intermediary with private industry and universities. A seven-member Executive Committee provided administrative control, appointing members to the four major technical committees and the various ad hoc special committees created to deal with specific problems as they arose.[3]

From the start, Hunsaker believed that scientific research was essential for the NACA's long-term viability, but he was fully cognizant, too, that forces threatened to pull the NACA away from fundamental research toward development and testing for industry and the military. The organization had recently come under scrutiny by the Brookings Institution, which recommended shutting down the committee and removing its functions to a new Department of Transportation. Although there was little sympathy in Congress for adding a new executive department, there were those who wanted to pare the NACA's budget, and the committee narrowly avoided a $100,000 reduction in its appropriation for the fiscal year 1939.[4]

In March 1939 Hunsaker assessed the status of aeronautical research in the United States, listing boundary-layer studies, flow separation, compressibility at high speeds, rocket propulsion, materials, flight control systems, weather forecasting, and radio communications as priorities. He stressed the importance of basic research, the benefits of which might not materialize for some time and that would not be attractive to private industry or the military:

> If I were Dictator, I would have a first-rate scientist, with necessary assistants and equipment, working continually in each of these fields of research, with no requirement as to a specific objective except to find out how and why and to publish the findings at intervals for the benefit of the other workers engaged on specific problems.

The German model, with generations of students working "under a gifted professor" such as Ludwig Prandtl, had been astonishingly productive. "In fact, practically the whole of modern aerodynamics, both theoretical and technical, can be attributed to Prandtl's 'School'. The American research effort of twenty years, while brilliant in discovering clever applications, has not made a single contribution to fundamental aerodynamic science."[5]

Hunsaker's ideas fell on the sympathetic ears of Vannevar Bush, chairman of the NACA Executive Committee since 1938 (replacing Joseph Ames, who had served in that capacity for nearly twenty-five years). In late June 1939, Bush appointed Hunsaker to chair an NACA committee to investigate means of coordinating the NACA and university research. Hunsaker called for a research coordinator, with authority over all aeronautical research at the NACA, the universities, and private industry. In contrast, George Lewis, one of the members of the Hunsaker panel, did not want to undermine his own considerable influence with industry and the NACA Main Committee in making decisions about research funding. By September Hunsaker and Lewis had agreed that the coordinator would work with industry and the universities to ensure that their programs were consistent with the guidelines established by the NACA technical subcommittees. The director of research would serve as an "executive officer" for the Main Committee and would not interfere with the work of the coordinator of research. Details of the compromise remained deliberately vague, leaving it up to Lewis and S. Paul Johnston, an MIT graduate who took over as the new coordinator of research in January 1940, to effect a working relationship.[6]

In the meantime, the nation held its collective breath while Europe blundered into its second war in a generation following Hitler's invasion of Poland in September 1939. Officially neutral, the United States prepared to bolster its defenses while providing moral and material aid to Britain and France. Hunsaker wrote to Philip Teed in January 1940, "Alice and I continue to read the papers and deplore the state of the world. . . . This country, as you know, is determined to stay out of the war, but the people seem to be equally determined to try to help 'save' the Democracies." While the press concentrated on Europe, Hunsaker worried about events on the other side of the Pacific, where Japan appeared to be on the brink of its own military adventurism.[7] Six months later, when the sudden fall of France left Britain alone to hold the line against Germany, the United States began to mobilize its scientific and technological resources for a conflict that looked inevitable.

One of the issues before Hunsaker and the NACA Aerodynamics Committee during these months of crisis was the location of a second facility to supplement the NACA Langley lab. The sentiment within the NACA was to locate the new laboratory on the West Coast—preferably at Sunnyvale, adjacent to Moffett Field south of San Francisco—but Theodore von Kármán and Robert A. Millikan at Caltech pushed for Pasadena. Hunsaker disagreed, believing that unless southern California industry was willing to help pay for wind tunnels and other expensive equipment at Caltech, the NACA should locate the laboratory at Sunnyvale. Charles A. Lindbergh, chairman of an ad hoc committee to investigate sites for the laboratory, came to St. Huberts in the summer of 1939 to discuss the matter with Hunsaker. The hero's visit created a stir in the little resort community; more significant, Hunsaker convinced Lindbergh that Sunnyvale was the best choice for the new laboratory. The site received approval from the NACA in the fall of 1939, and construction of the laboratory—named in honor of Joseph Ames, who was in poor health and would die in 1943—was well under way by the spring of 1940.[8]

Following the selection of Sunnyvale for a second aerodynamics laboratory was the NACA decision to establish a facility for aircraft power plant research. For some time, Hunsaker and Bush believed that the NACA needed to give more priority to propulsion research, reasoning that although much still remained to be done in aerodynamics—particularly in boundary-layer and turbulence studies—the most immediate performance benefits were most likely to come from improved aircraft engines. George Mead, formerly with United Aircraft and now chairman of the NACA Power Plants Committee, took the lead in planning for an engine laboratory, which received authorization and funding from Congress in June 1940. By the end of the year, a site adjacent to Cleveland's municipal airport had been chosen for the new facility, and construction began in January 1941.[9]

That month, Warner, burdened with his extra duties as a member of the Civil Aeronautics Board, asked Hunsaker if he was interested in the chairmanship of the Aerodynamics Committee. Quickly responding, Hunsaker said that he would be happy to take the new job, but he thought Bush needed to be aware that he was currently serving as a consultant to United Aircraft and American Export Airlines. He saw no reason to relinquish those consultancies, which "have given me no embarrassment," for in each case his services were strictly technical and did not extend to business or commercial matters. Bush concurred, and Hunsaker's appointment took effect on 1 February 1941.[10]

Like the NACA, the navy also struggled with the dichotomy between basic and applied research during the years before World War II. In the spring of 1941, the Naval Research Laboratory, the General Board, and the bureaus got into a turf battle over how to restructure the navy's research programs. To help resolve the matter, Secretary of the Navy Frank Knox, a veteran newspaperman who knew Hunsaker's father, called on Hunsaker. In a long memorandum on 7 April 1941, Hunsaker explained the distinction between the basic research of the Naval Research Laboratory and the developmental and experimental work done by corporations and navy bureaus. There was no time for top-to-bottom reform in a dangerous world, Hunsaker declared, "the development of a comprehensive and effective organization wholly within the Navy for scientific research would require the genius of an [Eli] Whitney and many years of growth." His recommendation was to bring in someone from outside the military—a "scientist of high professional prestige with wide contacts in the scientific world"—to coordinate navy research for the duration of the emergency. That person had to be entrusted with enough authority and independence to deal on an equal basis with his counterparts in the service.[11]

In his formal report to Undersecretary of the Navy James V. Forrestal on 27 June 1941, Hunsaker observed that there was a need for a "longer range effort to improve naval material through research and development" and suggested the establishment of a "central organization" to coordinate the bureaus' programs, create a research agenda, and establish lines of communication with other government agencies and private industry. He further recommended establishing a new Naval Research and Development Board in the Office of the Secretary of the Navy, chaired by a coordinator of research and development and including representatives from the Office of the Chief of Naval Operations and the bureaus of Aeronautics, Ships, Ordnance, and Yards and Docks. The coordinator was to be an eminent civilian scientist, with a naval officer as assistant. Hunsaker's proposal was clever. While recognizing the independence and prerogatives of the bureaus, it provided a foundation for enhancing the prestige and authority of the civilian scientific leadership, which he considered essential for the war effort. Knox liked the plan, and it appeared virtually unchanged as General Order 150 on 12 July 1941.[12]

It was imperative that the person selected as coordinator be thoroughly familiar with the navy's research establishment and have a sufficiently lofty scientific reputation to overcome resistance in the Naval Research Lab and the navy's technical bureaus. There was little question from the beginning that Knox wanted Hunsaker for the job. He agreed to take the position on 15 July, with the understanding that his service would be temporary and that he would be able to leave before the end of the year. Since the middle of March, Hunsaker had been spending much of his time in Washington in connection with his NACA responsibilities, so the new position did not change his routine much. There was some debate, though, about his status as a retired naval officer. Hunsaker had recently been promoted to reserve captain and could be recalled to active duty with that rank. Following discussions with Knox and Forrestal, however, he determined that as a captain he would have little influence on the flag officers in charge of the bureaus and decided he would be better off as a civilian reporting directly to the navy department's top leadership.[13]

Hunsaker lost no time organizing the coordinator's office. His staff consisted of a mix of regular officers and reserve personnel with scientific and engineering backgrounds. Hunsaker called them his "bird dogs" and indoctrinated them in how to get things done within the navy's Washington bureaucracy. The organization laid a firm foundation for the administration of naval scientific and technological research throughout the war. Nevertheless, Hunsaker's scheme depended heavily on the skills and influence of the person in the coordinator's office and did nothing to weaken the power of the bureaus, which continued to exercise their authority and independence in most matters relating to research. In 1945 he admitted to Rear Adm. Julius A. Furer, a friend and fellow Construction Corps officer who replaced him as coordinator of research on 16 December 1941, that despite everyone's best intentions "the degree of coordination and cooperation" among the bureaus during the war "has not been good."[14]

In addition to restructuring the navy's research organization in the summer of 1941, Hunsaker had definite ideas about how he would reorganize the NACA. From a small headquarters in 1939, with one laboratory and five hundred people, by the end of 1941 the organization had expanded its headquarters, added another laboratory with a third under construction, and nearly doubled the number of its personnel. Trying to avoid hurt feelings, he emphasized that none of his suggestions should be construed so "as to reflect unfavorably on individuals" but that many people in the NACA did not have a clear understanding of their respon-

sibilities. The upshot was that policies and programs were either not implemented or were overturned when key administrative personnel exercised veto
power. Compounding this were the NACA's "haphazard" planning and financial
procedures, adequate when the organization was smaller but now unworkable.

To fix things, Hunsaker recommended, first, drawing up an organization
chart with clear lines of responsibility and job descriptions, which he expected
would lessen interference from top-level personnel in the decision making and
implementation processes. Second, the NACA needed to "prepare complete
budgets," adopt standardized accounting procedures, and decentralize the system by transferring budgetary and planning authority to the laboratories. These
changes would achieve an equilibrium between the Main Committee and the
labs, allowing efficient decision making in Washington while preserving the
labs' "freedom of action" within the existing committee structure.[15]

Events moved rapidly over the next few months, culminating in Hunsaker
gaining more authority over the NACA than he could have imagined at the beginning of the year. Bush, who had taken on the chairmanship of the Main
Committee in addition to his duties as chairman of the Executive Committee,
agreed to become the director of the new Office of Scientific Research and Development in 1941. Wishing to limit his responsibilities, he canvassed members
of the NACA to see who they thought would be the best replacement for him as
chairman. Hunsaker was the obvious choice. Bush recalled, "we all thought
you were the logical man to head the thing up," and he was "delighted" that
Hunsaker was willing to succeed him.[16]

At the regular meeting of the NACA Executive Committee on 24 July, Bush announced his resignation, effective 31 July. The committee unanimously elected
Hunsaker to serve out the remainder of Bush's term. On accepting the position,
Hunsaker thanked those present "for their confidence in him," graciously
pledging to do his best "to measure up to the high standard set by the retiring
Chairman." At the NACA annual meeting on 23 October 1941—the first in
its new home at 1500 New Hampshire Avenue, where it had moved from its
cramped quarters in the Main Navy Building on Constitution Avenue—the
membership elected Hunsaker to full five-year terms as chairman of the Main
Committee and the Executive Committee.[17]

Hunsaker got his first taste of his new responsibilities on 3 October when he
argued the NACA case for advanced research before the House Appropriations
Committee. With billions of dollars in military aircraft procurement at stake,
Hunsaker told the committee, it was incumbent on Congress to make sure that
the NACA had the money, people, and facilities to do the job. He asked for and
received—in addition to $1.1 million to cover basic research—more than $7

million for the NACA fiscal year 1942 budget to pay for construction and equipment at the Langley, Ames, and Cleveland laboratories. He stressed that the army and navy depended on the NACA "to furnish the new knowledge—the result of both basic and applied research—that underlies and makes possible the advances that are being made in American aircraft today." Congress had to act now to ensure "a normal balance between basic and applied research" at the NACA to guarantee the acquisition of high-performance aircraft "in the event of a long war."[18]

Much of Hunsaker's attention in late 1941 focused on the two new laboratories. Ames was in good shape by December 1941, with many of the key personnel in place, three wind tunnels up and running, and the lab engaged in an aggressive research program. Cleveland, though, was another matter. There progress had been delayed by construction snarls, priority conflicts, and labor shortages. Hunsaker worked through George Mead to try to resolve some of the problems and to expedite completion of the laboratory. After Mead studied the situation in Cleveland, Hunsaker concluded in late November that the way to get things back on track was to bypass normal contracting procedures, which in turn meant more money from Congress.[19]

Less than two weeks later, everything at Cleveland and in the NACA took on more urgency. At MIT on Sunday, 7 December, Hunsaker was having lunch with Sir Richard V. Southwell, a British physicist and mathematician, when the stunning word came of the Japanese attack on Pearl Harbor. The next day Hunsaker was on the train to Washington, arriving to find the city in a frenzy as military and civilian officials sorted out the aftermath of the calamity and scrambled to find the best course of action for the nation at war. On the ninth, a day after Roosevelt's stirring address to Congress and the American declaration of war, Hunsaker conferred with NACA headquarters people before making his way up to Capitol Hill to plead for supplemental appropriations for the Cleveland laboratory.[20]

Hunsaker told the Senate Committee on Appropriations that work was already well along at Cleveland but that another $10 million had to be spent on the construction of the main laboratory building and the wind tunnel. He had received letters from Air Corps Chief of Staff Henry H. Arnold and Jack Towers, now a rear admiral and chief of the Bureau of Aeronautics, stressing the urgency of getting the Cleveland lab into operation and underscoring Hunsaker's and the NACA's decision to expedite construction by skipping the usual competitive bidding procedures and immediately signing cost-plus-fixed-fee contracts with qualified firms. Explaining that the NACA had already lined up potential contractors, Hunsaker added that none had expressed any interest because they had plenty of business and considered the undertaking too risky unless they were guaranteed a profit. Although there was some concern over the nature of the contracts, the committee agreed to the NACA request.[21]

The 1941 NACA annual report was Hunsaker's first. Submitted to the president only a week after Pearl Harbor, it emphasized aeronautics' key role in prosecuting and winning the war and the NACA research that had made production aircraft "more effective military weapons." Of immediate concern was maximizing the performance of American military airplanes, best accomplished, Hunsaker believed, by "cleaning up" their aerodynamics as well as by investigating new designs for wings, propellers, and engine cowlings. He continued to cling to his belief that it was essential for the NACA to maintain its independence from industry and the military and to focus on long-range research programs, although he conceded that for the duration of the war there would be intense pressure to concentrate on development and evaluation.[22]

The report paid close attention to the needs of the aviation industry, assuring manufacturers that the NACA was aware of their problems and would have representatives in their plants to discuss specific questions. Moreover, the NACA labs were open to all aircraft designers, who were free to consult with specialists about research that might be useful while their projects were still in the preliminary stages. But generating a cooperative atmosphere between the aircraft industry and the NACA was easier said than done. Writing in October 1941 in response to a misunderstanding about patents and the sharing of proprietary information with the NACA, Eugene E. Wilson, the president of United Aircraft and a former navy Construction Corps officer, said that although he understood the importance of "mutual confidence" between the manufacturers and the NACA, he doubted whether this was possible given the political nature of the committee and its members.[23]

During the first years of the war, Hunsaker instilled in the NACA the need to reassure the industry that it was on their side. In 1942 the NACA Western Coordination Office, created in 1940 at Sunnyvale, spun off a branch in Santa Monica specifically to work with manufacturers in the Los Angeles area. The Langley lab reported in 1943 that more than forty-five industry people were visiting the place daily. Changes in the NACA's publication policy also had a positive effect. The committee continued its usual practice of disseminating information through published technical reports and notes, supplemented after 1941 by additional publications at different levels of classification and informal notification of especially "hot" test results before they appeared in print.[24]

Hunsaker used his personal connections and friendships to keep NACA-industry relations cordial. In the summer of 1942, he wrote to Charles Lindbergh that wind tunnel work at the Langley lab would help improve the altitude performance of the B-24s Ford was manufacturing at Willow Run outside Detroit: "I am sure

Hunsaker, General George C. Marshall, and Orville Wright at an NACA luncheon, 1942. Hunsaker had become chairman of the NACA in the fall of 1941. (255-GF-392, Courtesy National Archives)

the NACA would be very glad, if you wish it, to send an engineer to Detroit with the details of our full-scale tunnel researches." He also visited aircraft factories and research facilities as often as his busy schedule permitted. In August 1942 he took the train to California to visit the Ames laboratory and Stanford University in the San Francisco area and the Douglas, North American, Lockheed, Northrop, Vultee, and Consolidated plants in Los Angeles and San Diego; a similar tour in the late winter of 1945 followed a nearly identical itinerary. During the war he also made trips to Akron, Dayton, and Detroit in connection with his duties as NACA chairman and as a director with Goodyear Tire and Rubber and the Chrysler Corporation.[25]

Committed as he was to science and fundamental research, Hunsaker soon realized that there was no way to avoid demands on the people and facilities of the NACA for more pedestrian business. After the war Hunsaker recalled that much of the NACA's work during the conflict involved "finding 'quick fixes' to make existing aircraft better performers and production engines more powerful." For engines, Hunsaker relied on Mead, who was of like mind that the NACA's concentration on testing and development work in support of industry was a necessary evil. Mead wished that the aircraft and engine companies had put in place the infrastructure necessary for their own research, freeing the NACA to concentrate, as he put it, on "more forward-looking work." Hunsaker had great respect for Mead's expertise, remembering him as a "perfectionist," who despite occasional ill health worked hard throughout the war.[26]

At a meeting of the National Academy of Sciences in Washington on 22 November 1943, Hunsaker outlined the NACA's power plant research, stressing the emphasis on extracting more energy from existing engines by improvements in supercharger design and the use of water injection and high-octane, antiknock fuels. The NACA assisted industry with new designs, among them the twenty-eight-cylinder, 3,000-horsepower Pratt and Whitney R-4360 radial engine slated for use in the next generation of long-range bombers. This was about the practical limit of the piston engine, Hunsaker believed; because of mechanical complexity, cooling problems, and consequent unreliability of such power plants, the "prospect for developing conventional piston engines giving more than 3,000-rated horsepower is not very good." On the other hand, the gas turbine—smaller, lighter, and mechanically simpler than the piston engine— "should eliminate a host of troubles inherent in the conventional engine," including vibration, cooling, and lubrication problems. In the form of the turbojet, the new technology held out the possibility of impressively higher speeds. Not that the gas turbine was without its own limitations—fuel economy being one—but Hunsaker was confident that advances in high-temperature materials would increase its efficiency.[27]

Indications early in the war were that the United States had fallen behind the British and Germans in jet propulsion, largely because the emphasis on ad hoc developmental programs had tied the NACA's hands in basic research. To outline the dimensions of the problem, the NACA had appointed a Committee on Jet Propulsion in March 1941, chaired by Stanford's eminent William F. Durand, with Hunsaker serving on the committee in an ex-officio status. Hunsaker recognized the advantages of jet propulsion as aircraft approached and exceeded 500 miles (805 kilometers) per hour, where propellers ran into compressibility problems. At the urging of the Army Air Forces (AAF), the Durand Committee recommended that Allis-Chalmers, Westinghouse, and General Electric be given contracts for advanced turbojet projects.[28]

United Aircraft was also involved in gas turbine work in 1941. Even though he had quit United's Technical Advisory Committee to avoid conflicts of interest, Hunsaker retained informal ties with the company, which allowed him access to its turbine research. Less than a week after Pearl Harbor, Leonard S. Hobbs of Pratt and Whitney wrote to Hunsaker to tell him that he had had "long discussions" with MIT's C. R. Soderberg about turbine engines and had recently met someone familiar with the British Gloster E.28/39, which had "quite reasonable" performance compared to conventionally powered aircraft. Hunsaker replied that Hobbs's "dope on the British development checks with mine" and thought that Pratt and Whitney's decision to go ahead with jet propulsion work was the "right track" to follow.[29]

Hunsaker's thinking did not square with that of the AAF, which had determined that such established aircraft engine firms as Pratt and Whitney should not be diverted from production of conventional engines and had decided that the best way to catch up with foreign developments was to acquire existing British technology. As a result, the AAF obtained one of Frank Whittle's centrifugal-flow turbojet engines from the British and assigned its manufacture to General Electric. Pressured by the service, Allis-Chalmers took on production of the de Havilland Goblin engine, which forced the company to abandon a promising axial-flow turbojet design, much to the dismay of company engineers.[30]

Writing in January 1945, Hunsaker thought it was still too early to tell whether the AAF's decision was a "mistake or a stroke of administrative genius," although he conceded that it meant the service got the engines it needed faster than if it had followed only the NACA's more deliberate, long-term development program.[31] The problem was that the AAF's decision to have General Electric and Allis-Chalmers go ahead with British turbojet designs left the impression that the NACA's emphasis on long-term research signified an unwillingness to meet the immediate demands of the military or to support the efforts of private industry to gain a competitive foothold in a potentially lucrative new technology.

Under such circumstances, Hunsaker had the unenviable task of guiding the NACA through a minefield of competing interests while trying to stay on top of the latest research on turbojets and related technologies. In December 1942 George Mead reported to Hunsaker that preliminary research had been promising and that the time was ripe to move into full-scale wind tunnel testing of gas turbines. Nevertheless, Mead agreed with Hunsaker that it was highly unlikely that work on jet propulsion would advance quickly enough to have much benefit during the war. It became obvious to Hunsaker by the middle of 1943 that as far as turbojet engines were concerned the military had lost confidence in the NACA and that the organization's perceived shortcomings had in part led to the AAF's decision to go it alone in this important field. Familiar with British advances in turbojet engines and some of the obstacles they had encountered with their design and construction, Hunsaker fretted that the military and the

British were not being as open as they could be in sharing turbojet information with the NACA.[32]

Leaving the NACA out of the loop in jet propulsion was not a consequence of internal deficiencies. Early on, the organization did its best to identify corporations to develop axial flow designs, only to have the AAF clamp on a lid of secrecy and move on its own to obtain British information and engines. Lost in the rhetoric and finger-pointing about the NACA's shortcomings was that the NACA was not supposed to invent anything but rather was to solve problems brought to its labs and to provide the resources that made invention possible. Through no fault of its own, the NACA found itself having to repair its tarnished image with both the military and the industry. It would not be easy, but the best way to do so was to demonstrate that it was willing to expedite the services' turbojet programs.

In keeping with this role, the new Cleveland laboratory, which after numerous delays had finally been dedicated in May 1943, began extensive testing of a General Electric Whittle-design centrifugal-flow turbojet before the year was out. Achingly prosaic, the work focused on endurance and altitude testing— hardly the cutting-edge experimentation that the NACA had been accustomed to. The Ames and Langley labs concentrated on research associated with gas flows in turbojets and the general aerodynamics of jet-propelled aircraft; they also continued to work on improved designs for aircraft engine supercharger compressors. Hunsaker offered the NACA's services and expertise to Bush's Office of Scientific Research and Development in the spring of 1944, as it set up a program to investigate the possibilities of ramjets for missile and aircraft propulsion.[33] These programs did nothing to boost the NACA's prestige in basic research, but they did contribute to the success of the turbojet program and kept the NACA involved, if only in the most circumscribed way.

In an attempt to highlight its involvement in the new technology and to mend fences with industry and the military, the NACA held a conference on jet propulsion in Washington in December 1943. Hunsaker, Durand, Lewis, and the Langley lab's Henry J. E. Reid and Eastman Jacobs met with representatives from industry and the military for a thoroughgoing appraisal of the latest developments in the field. The attendees went over the most recent British accomplishments and those of the Cleveland and Langley labs. Then Hunsaker reviewed the history of jet propulsion, surveying the characteristics of specific engines and underscoring how NACA research had stimulated the work of various companies and the military.[34] Still, he could do nothing to eradicate the impression that the NACA had been caught off guard by foreign developments and had fallen behind in this important technology. Frustrated, Hunsaker realized that he and the NACA were in a no-win situation, in which neither work in support of industry and the military nor basic aerodynamic research seemed enough to satisfy those critics who thought the organization should have done nothing less than invented the first turbojet-powered airplane.

Rocket technology was often linked to jet propulsion during the war. Having cut his teeth in aerodynamics, Hunsaker had only the vaguest interest in rockets, which seemed to him a crude, amateurish form of propulsion with little relevancy to aeronautics. Theodore von Kármán recalled that in 1938 the Army Air Corps had organized a special committee under the auspices of the National Academy of Sciences to outline key aeronautical research programs. Two pressing problems were windshield icing and visibility and rocket-assisted takeoff of overloaded aircraft. As MIT and Caltech agreed to divide up their research programs, Hunsaker told von Kármán that he was welcome to "take the Buck Rogers job" as long as MIT had priority over the cockpit visibility project.[35] Probably apocryphal and obviously intended to demonstrate Hunsaker's and MIT's short-sightedness, von Kármán's anecdote fails to convey the importance in the late 1930s of basic research into the phenomenon of icing or the practical limitations of rocket technology at the time.

Largely due to Hunsaker's lack of enthusiasm and the demands of other wartime projects, the NACA's direct involvement in rocket research was limited. As early as 1941, Hunsaker studied German work on rocket-assisted takeoff, which appeared to parallel the work at Caltech, and determined that the United States should at least monitor such developments. It also seemed possible to Hunsaker to use rockets to achieve greater speeds for short durations in combat, but few other applications made sense to him. He maintained a close liaison with Bush and the National Defense Research Committee, which had been formed in June 1940 and had responsibility for rockets and other programs under the auspices of the Office of Scientific Research and Development, and as a member of its council he stayed on top of all the latest weapons projects. Hunsaker also supported the NACA's establishment of a rocket research station at Wallops Island on the Virginia coast, largely because the work there assisted in understanding high-speed atmospheric flight.[36]

Jet and rocket propulsion directly affected the path the NACA took in basic aerodynamics research, a field in which Hunsaker felt more at home. As airplane speeds increased, the problem of compressibility arose. Up to the end of the 1930s, most aerodynamicists found it convenient to presume air to be an incompressible fluid, but as aircraft speeds exceeded 400 miles (644 kilometers) per hour—and as 500 miles (805 kilometers) per hour seemed attainable—it became apparent that compressibility effects might adversely influence aircraft control and stability. Hunsaker knew that the propeller was especially vulner-

Hunsaker at a meeting of the Office of Scientific Research and Development. His friend Vannevar Bush is in the front row, second from the left. (80G-59979, Courtesy National Archives)

able to compressibility, losing efficiency at high aircraft speeds, and advised that jet propulsion would bring additional aerodynamic problems.[37]

By 1944 it was clear that the aircraft production effort of the previous two years had paid off and that there was every expectation that the war would end in victory for the United States and its allies. For Hunsaker it was not too soon to give serious consideration to the direction of the NACA in the postwar years. Before the American Academy of Arts and Sciences in February he talked about the value of the NACA's aeronautical research as the "raw material" for the nation's wartime aviation industry. On the horizon, the turbojet could be expected to provide "very high speed" but at the cost of high fuel consumption and reduced range. In aerostructures new alloys and plastics promised greater strength and lower weight. The theoretical work of the NACA on low-drag wings and the effects of compressibility at high speeds had enhanced the combat effectiveness of American warplanes. In the future Hunsaker foresaw that "persistent research and experiment" would likely solve the control problems inherent in tailless and all-wing airplanes, which offered significant weight savings and drag reduction. In the years to come, extensive research was also certain to improve the stability and increase the payload capacity of helicopters.[38]

This second experience with world war meant Hunsaker was busier than he had been at almost any time in his life. Fifty-seven years old in the winter of 1944, he juggled teaching and administrative duties at MIT with navy, NACA, and other responsibilities in Washington. Usually he spent one or two days a week in Washington, taking an overnight train from Boston to minimize the working hours lost and staying at the Army and Navy Club off Farragut Square while he was in the city. The club's bar, he found, was a good place to relax and catch up on the latest Washington gossip. Occasional train trips to Langley, Cleveland, and Moffett Field combined inspection of burgeoning NACA and industry research facilities with meetings of the NACA Executive Committee.[39]

Hunsaker oversaw an almost exponential growth of the NACA during the war. In 1942, the first full year of the war, the organization expanded to nearly 1,800 people, followed by an increase of another 1,000 in 1943. By 1944 the NACA employed 4,500, and by 1945 had more than 6,000 on the payroll. Appropriations nearly doubled the second year of the war, from $11.2 million in 1941 to $19.9 million in 1942; by 1944 they reached $38.4 million, and in 1945 topped $40 million. Five committees and twenty-two subcommittees reported to the Main Committee.[40] No longer an intimate group of like-minded colleagues, the NACA had suddenly become a diversified bureaucracy with all the administrative and financial headaches that came with large organizations.

The war also brought changes to the Hunsaker family. At MIT, Peter earned his bachelor's degree in 1939 and, in 1941, a master's degree in mechanical engineering. His marriage to Frances Lowell on 27 June 1942, in Barnstable on Cape Cod, cemented the Hunsakers' ties to the old Boston aristocracy. Peter upheld the family tradition by joining the navy and serving as an officer during the war. Meanwhile, Jerry Jr. received his bachelor's degree from Harvard and went on to earn a master's degree in business administration in 1940 from Harvard Business School. He got a position with Arthur Anderson and Company in New York and joined the navy as a reserve ensign before Pearl Harbor. Jerry became engaged in November 1942 to Elizabeth Blake. Ever the engineer, Hunsaker proudly told George Mead that "Jerry did a good job of picking" his bride to be, the granddaughter of Bostonian Francis Blake, who had invented the transmitter used by the Bell Company in its first telephones. The wedding took place on 8 May 1943 in Weston, Massachusetts. Shortly after leaving Vassar, young Alice fell in love with a much-decorated British army major and architecture student, Tommy Bird, who was in the United States with the British military mission in Washington.[41]

Through MIT, through Bush and the National Defense Research Committee, as well as through his NACA connections, Hunsaker found himself in the middle of a wartime project to develop radio-controlled guided bombs. Known as AZON, an acronym for guidance in "azimuth only," the bomb program—begun in September 1940 by the National Defense Research Committee—was a joint effort by universities, industry, and the AAF. Two MIT aeronautical engineers, L. O. Grondahl and Julius P. Molnar, saw the possibility of guiding a bomb to its target using a compact television camera and radio signals from an airplane in flight. They recommended to Bush that the National Defense Research Committee begin a study of television, radio, and the aerodynamics of freefalling bombs and in the winter of 1940–41 began their own work on bomb aerodynamics.[42]

The bomb project got under way in early 1941. Collaborating with the MIT engineers were two Pittsburgh firms, Gulf Research and Development Company and the Union Switch and Signal Company (where Grondahl had become director of research). Using the Wright Brothers Memorial Wind Tunnel, the researchers learned that bombs could be maneuvered up to a thousand feet (305 meters) from side to side in the final moments before impact. They decided to employ a guidance system controlling the bomb in roll with a gyroscope and using direct sight from the airplane and radio guidance to control its movement laterally, or in azimuth. Hunsaker suggested changes in the design of the bomb's control surfaces and assured Grondahl that he had the full backing of the Aeronautical Engineering Department at MIT and access to the research facilities at the NACA Langley lab. It turned out that the press of other business prevented MIT from playing a larger role in AZON research, and Gulf assumed most of the responsibility for the project after 1942. The AZON bombs saw limited use in the China-Burma-India theater in 1944. After the war, Hunsaker praised Grondahl for making "an outstanding success of a very tricky development puzzle" in bringing AZON to fruition.[43]

With the war winding down in 1945, Hunsaker anticipated that the end of the conflict would permit the reestablishment of the international research community, which he believed was crucial to technical progress. He and others in aeronautics saw the International Union of Theoretical and Applied Mechanics as one means of restoring the scientific and technological discourse disrupted by the war. His connection to the organization went back to 1924, when he attended the first congress in Delft, Holland, as a representative of the NACA. He went on to help organize the fifth meeting of the union at MIT in 1938, which in addition to the dedication of the Wright Brothers Memorial Wind Tunnel featured a symposium on turbulence organized by Ludwig Prandtl and presentations by other luminaries in aeronautical engineering. Plans to hold the sixth congress in Paris in 1942 collapsed with the outbreak of the war. In November 1944 Hunsaker wrote to a friend at the École Polytechnique in Paris who had thankfully "survived the long dark night" of German occupation, telling him that "we live in the hope that we can restore the kind of world that makes such meetings again possible." The Paris meeting finally came off in September 1946, although Hunsaker was unable to attend.[44]

Thinking about the postwar world forced Hunsaker to direct his attention to

Hunsaker and representatives of the East Coast aircraft industry at the NACA's Cleveland lab, May 1944. Hunsaker had a difficult balancing act, seeking industry support while preserving the NACA's independence. (255-GF-102, Courtesy National Archives)

the NACA's future research. To ascertain what aircraft and engine manufacturers wanted, Hunsaker met with representatives of the East Coast aircraft industry at the Cleveland lab in May 1944 and with their West Coast counterparts at the Ames lab a month later. At the meetings there was "general agreement" that the NACA should disseminate the results of its research to American companies before it went to their European competitors, that it should perform testing and evaluation as needed by the government, and finally, that it should not engage in research that overlapped that of industry in such fields as metallurgy, electronics, and fuels. Disagreement remained about how much research money the NACA should get, the testing and evaluation of specific products, which might give an advantage to companies with smaller engineering staffs, and the status of the Cleveland lab, regarded as a "potential threat" to the aircraft engine industry.[45]

At the 27 July 1944 meeting of the NACA Executive Committee, discussions centered on Hunsaker's reports on postwar research policy. Everyone understood the industry's concerns about government competition in what was likely to be a severely contracted business after the war; as Hunsaker put it, "the industry feels that it is faced with the problem of survival." Assurances that the

NACA would limit its activities to research, while leaving design, development, and testing to private firms, would ease some of industry's worries. On the other hand, Hunsaker got the impression from the meetings in Cleveland that the industry would only be satisfied if the NACA ceased all power plant research. Some manufacturers thought "the NACA should fold up" its engine research, Hunsaker explained, which was not a realistic possibility given the institutional commitment and financial investment the government had made at the Cleveland lab. In Hunsaker's opinion, industry missed the point. Facing a potential challenge in the air from the British in the postwar years, it was in the country's interest to formulate a unified policy of government-industry-military cooperation in aeronautical research.[46]

Before the year was out, Hunsaker provided Vannevar Bush with thoughts on specific areas of research for the NACA after the war. He was especially interested in guided missiles, a military program for which the NACA would likely be called upon to support. Much of the work that had been done at the Langley and Ames labs, Hunsaker believed, had been little more than "bug chasing" and had not maximized the labs' potential. The NACA research facilities could do much with high-speed aerodynamics, stability and control, materials, and turbojet engines. Its supersonic research, in particular, was directly applicable to high-performance guided missiles. Its labs had four supersonic wind tunnels; and the labs had drop tested models at speeds up to Mach 1.1 and had fired objects through an evacuated tube to record flow data at supersonic speeds. The Cleveland lab's investigation of internal flows resulted in improved intakes and ducting to maximize turbojet engine performance. Hunsaker considered the NACA's relationship to the military in the development of guided missiles "analogous" to that for aircraft development and offered the services of the NACA to Bush and the National Defense Research Committee in fostering guided missile research.[47]

An opportunity to make his case for redirecting the NACA's research agenda and policy came when Hunsaker testified before the House Select Committee on Post-War Military Policy, chaired by Democratic Congressman Clifton A. Woodrum of Virginia. Speaking to the committee on 26 January 1945, Hunsaker saw aeronautics "entering an era of revolutionary change" through jet and rocket propulsion, which offered "extraordinary high-speed possibilities," potentially in the supersonic range. In this challenging new arena, it was incumbent on the government to use the NACA's "special research facilities and personnel to provide our designers with information permitting the industry to build the kind of airplanes and engines that the United States must have to maintain its superior position in both military and civil aviation." Not to do so was risky,

Hunsaker and Fleet Admiral Ernest J. King at a navy research conference at the NACA's Cleveland lab, July 1945. Hunsaker retained close ties to the navy's top leadership throughout his professional career. (255-GF-NACA C-11086, Courtesy National Archives)

because American technical leadership could easily be "lost in war to an alert enemy, and in the peace to come, to a vigorous commercial competitor."[48]

In general, Hunsaker preferred conservative solutions to these revolutionary problems. He did not want to see any major change in the NACA's diversified committee system, which provided the flexibility to address research problems for both private industry and the military. To merge the NACA into a unified Department of Defense, as some suggested, would be a "disaster"; it would "subordinate" civil aeronautics to military requirements, and the inevitable secrecy would stifle the flow of information. At the same time, the NACA could not give the appearance of competing with private industry. It would not, Hunsaker insisted, be involved in the development of specific aircraft, nor would it apply "research results to the design of improved aircraft and equipment." Its research labs were to serve industry, with timely dissemination of the results of their work. It would continue to contract out research to universities and to government agencies with expertise in specific areas, and it would emphasize coordination and cooperation to avoid duplication of effort. Members of the subcommittee praised the NACA for its "wonderful work" during the war, rewarding the organization with $4.5 million in navy funds for supersonic research.[49]

In December 1944 President Roosevelt appointed Hunsaker to the NACA Main Committee for another five-year term. The reappointment demonstrated not only that the administration valued Hunsaker's leadership and wise counsel during the war but that it wished to see his contribution continue into the postwar years.[50] As much as anyone in the expanding military-industrial-university complex, he saw what lay ahead for the NACA in an uncertain world. Hunsaker knew that it was one thing to proclaim the NACA's commitment to basic research in the exciting new fields of aerodynamics, materials, and propulsion but that it was something else again to make those proclamations reality, given reduced budgets and competing constituencies. He knew also that peacetime administrative and technological threats to the NACA would be harder to counter than the problems caused by war. Peace presented difficult and controversial choices, which would challenge Hunsaker's technical ability, organizational skills, and political expertise.

10.   New Directions and Old

N ews of the Japanese surrender on 14 August 1945 brought a shared sense of relief and a feeling of a job well done among the NACA's more than six thousand people scattered across the country. They had worked long and hard to ensure that America's fighting forces flew the best possible airplanes and benefited from the most up-to-date aviation technology. Before the war it had been generally assumed that most of the NACA's work involved basic research, but that changed when the organization, out of expediency, was thrown into the breach as a development and evaluation agency for the military. Jerome Hunsaker, as NACA chairman, was uneasy about the imbalance in the organization's responsibilities, knowing full well that it had to be realistic about its wartime obligations. Now, at the end of the conflict and with revolutionary technological changes taking place at a dizzying rate, it was clear to him that the NACA would have to traverse a political, fiscal, and technological thicket to regain its standing as one of the nation's premier research institutions while not alienating industry or the military.

Hunsaker's first opportunity to demonstrate publicly where he thought the NACA should be headed came in February 1946, when he testified before a special Senate committee investigating the national defense program. Chaired by Democrat James M. Mead of New York, the committee accumulated thousands of pages of testimony relating to the structure of the military, the size and status of the aircraft industry, and the responsibilities of governmental research agencies in the postwar era. In the parade of industry representatives and military officers who marched before the committee were those who claimed that, largely due to the shortcomings of the NACA and its leadership, Germany had achieved superiority in such key technologies as jet and rocket propulsion and high-speed aerodynamics. When he appeared before the committee on 27 February, Hunsaker had to fight a rear-guard action to defend the NACA's record during the war while at the same time stressing the organization's commitment to future basic and applied aeronautical research.[1]

Hunsaker dismissed accusations that the NACA had not been cooperative in sharing the results of its latest research with the aircraft industry, explaining that the organization had "quite an elaborate setup" to ensure a smooth flow of information and to meet the needs of the NACA's primary constituencies. In June 1945 the NACA had established the Industry Consulting Committee, charged with advising it on research and facilities in the postwar years. Additionally, seventy-one members, or about 25 percent, of those on NACA subcommittees represented the aircraft industry, more than enough to ensure access to data and to give direction to the organization's research programs. Care had always been taken to verify the results of the NACA's research and to make it available to the industry at the earliest possible time. The low-drag, laminar-flow airfoil used in the North American P-51 Mustang fighter was only one example of how the system worked to the benefit of industry. At the same time, Hunsaker cautioned that the NACA could not side with aircraft manufacturers to the point of jeopardizing its independence or the public interest.[2]

In answer to allegations that the Germans had devoted more personnel and money to research than had the NACA, Hunsaker pointed out that much of Germany's "scientific" effort had in reality been engineering aimed at the deployment of rockets—"weapons of despair" that "diverted a lot of their industry." He did admit the NACA could have done more, but there were practical limits to the gathering of information in wartime. He said that he personally knew what "any skilled person ought to know" in his profession but did not think he or the NACA had "responsibility for conducting any espionage organization anywhere." Accusing the NACA of failing to devote adequate attention to rocket or jet propulsion was like condemning the entire medical profession for not finding a cure for cancer after it chose to focus on other public health problems. "The decision was made, I think, with our eyes open, that we were in a war. The duty of the research men then was to support the procurement program." After all, he reminded the committee, "we won the war."[3]

Much of Hunsaker's time before the committee was spent discussing money. Had Congress been more generous in making funds available to the NACA during the 1930s, he argued, the organization might have been able to take a more aggressive approach to research, particularly in some of the areas that it was now being accused of not pursuing. Advanced research cost money and demanded skilled people; unless Congress made a commitment to both, the country ran the risk of not keeping pace. At the same time, he cautioned that it was more than just a matter of throwing money at the problem; thorough planning was needed, and it made no sense to appropriate large sums of money for a project one year and nothing the next.[4]

In a paper presented to the National Academy of Sciences in April 1946, Hunsaker was more specific about the NACA's propulsion research. For some time, engineers had known that, due to compressibility effects, the conventional piston-propeller combination fell off in efficiency at speeds approaching that of sound. The turbojet engine, despite its deficiencies, offered the ultimate solution to high-speed flight. On the other hand, the rocket, which had to carry its own oxidant, suffered an unacceptable weight penalty in comparison to other means of propulsion. "Since the energy contents of available propellants are not very different, rockets of much greater power are not in sight," he predicted, with the result that the rocket was likely to be limited to short ranges. Spectacular as it was, the German wartime rocket program had consumed scarce resources, "which the Germans could ill afford." "One is tempted to speculate about the possibilities of an improved rocket of this [V-2] type," Hunsaker went on. "An engineer cannot see much prospect for an improved propellant or much better materials of construction. . . . It therefore appears that the range of 200 miles is near the maximum for the type."[5]

Hunsaker's carefully reasoned analysis of the rocket misled him in the same way it did his esteemed colleague, Vannevar Bush, who in December 1945 had flatly declared that the intercontinental-range ballistic missile was an impossibility.[6] As he had been with the rigid airship, Hunsaker was locked in to a conventional way of thinking that made it hard for him to look beyond familiar practices and embrace reasonable technological alternatives.

In addition to money, Hunsaker had to wrestle with the question of the appropriate role of private industry in the NACA's governance structure. More than anything else, Hunsaker cherished the organization's autonomy and independence. Even before the war ended, aircraft and engine manufacturers had begun to agitate for a louder voice in the NACA, preferably through permanent representation on the Main Committee. Hunsaker hoped the creation of the Industry Consulting Committee would mute some of the most strident complaints, but the issue did not go away. In December 1946 Bush wrote to Hunsaker that he had "no doubt" that a person from the aircraft industry could "divorce himself in his thinking from his industrial connections to sit on a public board as a representative of his profession," but he did not believe that either Congress or the public could be convinced that that was possible. Enlarging the Main Committee to allow direct industry representation would make that committee unmanageably large and might open the door to other special interest groups, such as pilots, to request membership.[7]

Bush's letter took some time for Hunsaker to digest. He finally replied on Christmas Eve to tell Bush how much he valued his "counsel as the NACA becomes involved in new fields of applied science and new areas of public interest" and that he appreciated his concerns about the organization's independence. "Conflict of interest or misunderstanding could, in my opinion, never arise," because committee appointments with industry connections—including Hunsaker himself, who was on the boards of companies with aviation interests—had been made with "confidence in our ability to serve in an individual professional capacity" regardless of who one worked for. Complete autonomy was neither possible nor desirable, for the skills and knowledge that made many NACA members attractive in the first place stemmed of necessity from their work in the private sector. Nevertheless, he agreed with Bush that permanent industry representation on the Main Committee was not desirable.[8]

Little had changed as far as Hunsaker was concerned when he wrote in November 1948 to H. Mansfield Horner, chairman of the NACA's Industry Consulting Committee, about the issue of industry representation. Hunsaker acknowledged that members of the NACA whose experience had been gained in industry naturally had to have "knowledge of and be able to discuss the work interests and needs of the industry," but they did not in any way represent their employers. He went on to say that federal expenditures for aeronautics were best administered by independent-minded people who held the public interest sacred and that a "representative" committee would be mired in endless political debates. Hunsaker trusted the "good judgment and professional integrity" of members outside the government to make the NACA's committee system function for the public good.[9]

Hunsaker's trust in the integrity of public-spirited professionals to work in the best interests of the NACA and the nation reflected a conviction he shared with Vannevar Bush: that the technical expert was the person best suited to ensure the well-being of the nation in a dangerous new world. No technocratic conspiracy or intent to usurp traditional democratic values lay behind this determination. On the other hand, Hunsaker intuitively suspected that placing such faith in the rationality, independence, and good will of his allies in the business world was fraught with peril, and he strenuously resisted having one of them appointed to a permanent seat on the Main Committee.

Another concern besides NACA relations with industry was setting priorities for postwar research, one of the most important of which was high-speed aerodynamics. With the end of the war in sight, Hunsaker recognized that work with advanced guided missiles would demand new and costly aerodynamic research facilities. In December 1944 Hunsaker had prepared a long memorandum in which he outlined the NACA's role in air-breathing guided missile research, specifically emphasizing the organization's expertise in all aspects of aerodynamics, materials, and propulsion systems.[10]

To criticism from the Mead committee that at the end of the war Germany was three to four years ahead of the United States in supersonic research, Hunsaker responded that the Germans had done little research at high Mach numbers and that in some respects the NACA had done better in the vital area of transonic research. For research at higher speeds—beyond the capability of current wind tunnels—the NACA had begun high-altitude drop tests and had used rockets to investigate stability and control at supersonic velocities. There was still much to be done: Hunsaker called on Congress to support the creation of new facilities to explore high-speed aerodynamics in conjunction with guided missile and advanced propulsion research.[11]

Foreign developments in high-speed aerodynamics also pushed Hunsaker and the NACA to formulate policies and priorities for the postwar years. The Alsos Mission, undertaken immediately after the German surrender, yielded a tremendous quantity of information about German research in high-speed aerodynamics. From his post–World War I experience and knowledge of German research, Hunsaker recognized that the NACA had to obtain a great deal of information quickly, although he cautioned against committing the NACA to a line of research simply because the Germans had pursued it during the war. Representatives of the NACA on the Alsos Mission discovered that the Germans had begun construction of a 100,000-horsepower supersonic tunnel and were in the preliminary planning stages of building an even bigger (500,000-horsepower) model. Nothing in the United States had such capacity or flexibility. Hunsaker well knew that size was all-important in these high-speed tunnels, in which supersonic shock waves rebounded from the walls of the test section and created the phenomenon known as "choking." When in the fall of 1945 the Army Air Forces began formulating a program for building advanced wind tunnels, Hunsaker took the initiative of proposing a "unitary plan" in which the NACA and the military would, together, develop new supersonic research facilities.[12]

The idea was appealing to Hunsaker for several reasons. A unitary plan would make the best use of the taxpayer's dollars, which might be in short supply after the war, and it would ensure a continued NACA role in the latest aerodynamic research while simultaneously blunting criticism that the organization had not been cooperative in sharing information with industry and the military. Furthermore, if handled properly, the plan would clearly divide the research effort, apportioning to the NACA primary responsibility for basic research and leaving the development and testing work to industry and the military. It appeared, at least initially, to be a win-win situation for the NACA.

Hunsaker's ideas were consistent with the NACA's contribution to the national aeronautical research policy approved on 21 March 1946. Drafted by a special NACA committee and agreed to by the NACA, the army, the navy, and the Civil Aeronautics Administration, the document delineated and codified the responsibilities of government agencies and the aircraft industry in planning and coordinating research. Particularly gratifying to Hunsaker was the statement that the public interest demanded a commitment to modernizing older aeronautical research facilities and building new ones as needed. Most important, the document included wording that the NACA's "principal objective" was fundamental research in the aeronautical sciences, with the goal of solving the basic problems of flight. Application of the results of the NACA's research and development and testing, on the other hand, were left to the military and industry. It was essential to maintain these lines of responsibility to avoid "unnecessary duplication of facilities and effort" among the various parties to the understanding.[13]

Much to Hunsaker's chagrin, the policy, which should have resolved once and for all the NACA's primacy in scientific research, did not deter the AAF from pushing ahead with grandiose programs that trespassed on the NACA's "turf." In response to the Army Air Forces' insistence on having its own high-speed wind tunnel program, the NACA Main Committee at its semiannual meeting on 25 April 1946 authorized Hunsaker to appoint a committee headed by Arthur E. Raymond, chief engineer for Douglas Aircraft, to study the feasibility of a national supersonic laboratory. Among the other members of the committee were George Lewis and Hugh Dryden of the National Bureau of Standards, who chaired the NACA Subcommittee on High-Speed Aerodynamics. The Raymond committee met for the first time on May 10.[14]

Within a month the Raymond committee had a proposal, which Hunsaker summarized at the meeting of the NACA Executive Committee on 6 June. Included in the recommendations were a National Supersonic Research Center operated by the NACA, an engineering research facility for the AAF, and moderate-sized wind tunnels for selected universities. The plan adhered to Hunsaker's ideas about the NACA's preeminence in basic research, clearly delineating the military's developmental role and spelling out the lines of demarcation he had wanted since coming to the NACA chairmanship in 1941. Following up on the Raymond committee's proposal, the NACA created a special supersonic facilities committee chaired by Hunsaker and including Rear Adm. Lawrence B. Richardson and AAF Maj. Gen. Edward M. Powers. After a series of meetings at Wright Field, with Langley engineer John Stack representing the NACA, the AAF hired the St. Louis engineering firm of Sverdrup and Parcel to begin studies of a cooperative wind tunnel program.[15]

In the meantime, Hunsaker had left for England. The official reason for the trip was to attend the Isaac Newton Tercentenary celebration at Cambridge University, 15–20 July 1946, as a representative of the National Academy of Sciences, but he also saw it as an opportunity to learn where the British and others were headed with their postwar aerodynamic programs. Although regular transatlantic air services were now a reality, he and Alice made the crossing by sea, sailing from New York on 20 June on the Cunard liner *Queen Mary*. In London, the Hunsakers spent a few days with their daughter Alice, recently married and living with her British husband in the city's Kensington district.[16]

In his Cambridge presentation, Hunsaker stressed the NACA's contributions to aerodynamic research, especially in laminar flow and transonic flight. The development of a laminar-flow wing during the war owed much to basic research in fluid mechanics and to studies by NACA engineers on the control of air in the critical boundary layer close to an aerodynamic surface. Transonic flight was in many respects more challenging than supersonic flight, creating disruption in flow patterns as one section of the aircraft's wing attained supersonic speeds before other sections. To make matters worse, the flow of air in the wake of the wing at transonic speeds fluctuated to such an extent that it generated buffeting in the aircraft's tail surfaces, leading to instability and control difficulties. There was also the problem of compressibility when shock waves formed in front of and along the fuselage of an airplane as it approached and exceeded supersonic speeds, upsetting the flow of air in the boundary layer near the surface of the wing and drastically altering the wing's lift.[17]

His Cambridge responsibilities out of the way, Hunsaker was free to concentrate on aerodynamics and propulsion. Unable to get to Germany, he reported that the British already had employed a few of the top German aerodynamicists and were in the process of obtaining some of the power equipment and instruments from one of the large German wind tunnels. He spent two days visiting the Royal Aircraft Establishment at Farnborough, looking at structures experiments, boundary-layer research, and discussing the RAE's interest in building a big transonic tunnel. Afterward, Hunsaker toured aircraft and engine plants. At de Havilland he got a close look at the new jet-powered DH.108 Swallow fighter, which featured a "tailless" design and was capable of speeds approaching that of sound. Rolls Royce was well along with two types of production turbojet engines. Despite materials problems and severe limits on the engines' number of operational hours, Hunsaker believed that "Rolls has gone much farther than any other builder" and that British "momentum is strong" in the field. After an enjoyable and productive stay in England, the Hunsakers sailed for home on 22 July aboard the old *Aquitania,* which "fairly swarmed with troops, brides, and babies." They finally arrived back in Boston by way of Halifax and Montreal.[18]

Hunsaker's trip to England went smoothly in large measure because of the arrangements made by John Jay Ide, who had headed the NACA Paris office before it closed in 1940 and who since 1944 had been serving as a naval air attaché in London. In response to Ide's query in August 1945 about reopening the Paris office, Hunsaker had written that "there is no prospect of this for the foreseeable future." As much as he liked Ide, Hunsaker was not so sure that he was capable of absorbing the latest technical information. He confided to Lewis that "Ide is 100 per cent in arranging schedules, 10 per cent on understanding what is going on. He is liked and could be helpful to us as head of an office to arrange swaps of people, visits, etc." Moreover, Ide's responsibilities overlapped the intelligence-gathering activities of the military services. Hunsaker still believed that the NACA needed a permanent representative abroad to keep abreast of the latest technical developments and thought it possible for the NACA to coordinate all aeronautical intelligence from Europe provided that its representative were placed high enough and had the necessary technical background. Nothing happened until 1949, when the NACA finally received funding from Congress to reestablish the European office in Paris, with Ide again in charge.[19]

There was also the unitary wind tunnel plan, now threatened, Hunsaker believed, by demands by the AAF for unnecessary and exorbitantly expensive facilities. At a meeting of the supersonic facilities committee on 16 August 1946, AAF representatives pushed hard for approval of three large wind tunnels for testing full-scale turbojet engines and large engine-propeller combinations. Hunsaker objected. To him it was more logical to study parts of the aerodynamic flow problems in jet engines by using scale models in smaller and more economical tunnels. When the committee reconvened in October to discuss details of the unitary plan, the AAF continued to press for the big tunnels. Consulting engineers Sverdrup and Parcel estimated the cost of the NACA's National Supersonic Research Center at about $800 million and the AAF Air Engineering Development Center (AEDC) at Tullahoma, Tennessee—with three large forty-by-forty-foot (12.2-by-12.2-meter) transonic wind tunnels—at $1.8 billion. Staggering figures, totaling more than that spent on the wartime Manhattan Project to develop the atomic bomb, they were hardly likely to attract support in Congress. Hunsaker preferred a more modest start to the program by creating priorities, although the military people thought stretching the program out over eight to ten years was a better way to minimize the annual budgetary strains without jeopardizing any single component in the plan.[20]

Hunsaker persuaded the committee to agree that the unitary plan should give top priority to a single twenty-by-twenty-foot (6.1-by-6.1-meter) transonic propulsion tunnel with an innovative open-throat design, replacing one of the AAF's proposed forty-by-forty-foot (12.2-by-12.2-meter) tunnels at the AEDC. Two eight-by-eight-foot (2.4-by-2.4-meter) supersonic tunnels for both aerodynamic and propulsion research would be installed at the Langley and Cleveland labs, with another two going to the Ames lab, on the West Coast. Two smaller tunnels (two-by-two-foot [0.6-by-0.6-meter]) would be built at Langley and at the navy's David W. Taylor Model Basin in Carderock, Maryland. Powers eventually got Hunsaker to agree to only one eight-by-eight-foot (2.4-by-2.4-meter) tunnel at Ames and conceded that a single forty-by-forty-foot (12.2-by-12.2-meter) propulsion tunnel would be sufficient for the AEDC, provided it was pressurized for altitude research and was equipped to test full-scale engine-propeller combinations.[21]

No one, least of all Hunsaker, was completely happy with the compromise. His major concern was that the line of demarcation between facilities earmarked for research and those set aside for development had become blurred. Nevertheless, confident that he had a workable plan, Hunsaker forwarded the committee's recommendations to the NACA Executive Committee for consideration on 24 October. He emphasized that the price tag for the unitary program was on the order of $2 billion, with total annual operating costs somewhere in the neighborhood of $140 million—still a lot of money, but initially "only" $750 million would be needed to get started with the top priority items in the program. Vannevar Bush was "frankly appalled at the size of the proposition" and worried about its political implications, but the NACA Executive Committee unanimously agreed to pass the plan along to the Joint Research and Development Board for its approval before submitting it to the Bureau of the Budget. Because Bush chaired the Joint Research and Development Board and Hunsaker was a member, they anticipated no problems.[22]

The Hunsaker committee met again on 19 December to discuss the proposed tunnel for the AEDC. Hunsaker and John Stack questioned whether the tunnel needed the range of altitudes, speeds, and temperatures that the AAF wanted, hinting that the smaller, twenty-by-twenty-foot (6.1-by-6.1-meter), open-throat tunnel might work just as well. When General Powers insisted that the tunnel be capable of speeds up to Mach 1.6, Hunsaker suggested that the facility be built for subsonic speeds with provisions in the design for its conversion to higher velocities at a later date. Powers was not happy, but he went along with the compromise, provided the big tunnel was still given the same high priority agreed to in the October meetings. The next month, the NACA approved the unitary plan substantially as outlined by the Hunsaker committee.[23]

Wind tunnels were only part of the NACA effort to find solutions to the problems of high-speed flight in the early postwar years. To validate laboratory tests, Hunsaker thought it was necessary to fly aircraft through the entire speed range and into the unknown of supersonics, which required a series of high-performance test aircraft. Beginning in the fall of 1946, the NACA initiated a series of tests with the rocket-powered Bell XS-1, aimed at exploring the unknowns of high-speed flight. The site was the AAF base at remote Muroc, California, where the NACA set up a flight test unit under the administration of its Langley lab. A milestone in the program came in October 1947 when the XS-1 became the first manned airplane to exceed the speed of sound in level flight. Meanwhile, the jet-powered Douglas D-558-1 allowed extended flights at speeds nearing Mach 0.9, returning valuable data on aircraft stability in the transonic regime.[24]

While he was laboring with the unitary plan, Hunsaker faced a difficult personnel decision at the highest levels of the NACA staff. The highly capable NACA director of research, George Lewis, his health failing in late 1946 and early 1947, decided that he would have to curtail some of his activities. Lewis had put in long hours in exemplary service during the war, which more than likely contributed to his physical deterioration. Charged with setting the NACA's research agenda, Lewis understood that the organization often had to address the immediate technical problems facing the American aircraft industry. Disagree as he might with Lewis on the relative merits of basic research, Hunsaker had the utmost respect for Lewis's professionalism, and the two got on well together during the war years. At the 11 August 1947 meeting of the Executive Committee Lewis announced his resignation effective 1 September. Afterward, he stayed on as a research consultant; only days before his death from a stroke in July 1948, he wrote Hunsaker about recent aircraft engine research by Pratt and Whitney. Hunsaker regretted losing Lewis and welcomed naming the Cleveland lab in his memory, but he also knew that he had an opportunity to replace him with someone whose ideas were more in tune with his own on restoring the equilibrium between basic and applied research.[25]

Lewis recommended Hugh L. Dryden, associate director of the National Bureau of Standards, to succeed him. Hunsaker readily concurred. Dryden had done cutting-edge research on wind tunnel turbulence and boundary layers and had had close connections with the NACA for years, culminating in membership on its Aerodynamics Committee. Hunsaker had known Dryden since the early 1930s through the Institute of the Aeronautical Sciences and had worked with him on the NACA; the two had become good friends. He was most impressed by Dryden's intelligence, scientific background, administrative experience, and unwavering commitment to basic research. Dryden may have lacked Lewis's attention for detail and political shrewdness, but with his superior scientific background he was exactly who Hunsaker sought to take the NACA into new areas of research.[26]

From early 1947 on, the unitary plan passed beyond Hunsaker's control, although he continued efforts to secure its earliest possible implementation. A major obstacle in the spring of 1947 was an AAF proposal for a large hypersonic wind tunnel for missile research at the AEDC. Hypersonics—Mach 5 and beyond—was a potentially important area of basic research, involving more than just the aerodynamic problems associated with high-speed flight. At such velocities an enormous quantity of heat was generated by air compression and by the shock waves that accompany flight in the hypersonic regime. Engineers now had to apply to their research what they knew about "normal" fluid flows as well as analyzing a whole range of theoretical information about thermodynamics, heat transfer, and the properties of exotic materials capable of withstanding such extreme temperatures. The only glimpse of this new realm was what the Germans had encountered with their V-2 ballistic missile, which flew at speeds up to 3,400 miles (5,472 kilometers) per hour. At the end of the war the Germans were building a wind tunnel capable of producing speeds up to Mach 10. To explore this perplexing dimension of flight involving basic research and heavy doses of theory, the NACA had under construction at Langley an eleven-by-eleven-inch (twenty-eight-by-twenty-eight-centimeter) wind tunnel that could generate speeds up to Mach 7.[27]

It was obvious to Hunsaker that the AAF proposal violated the spirit of the unitary plan in going beyond the bounds of developmental work, that it was a needless duplication of the Langley tunnel, and that it infringed on the NACA's role in basic research. Moreover, largely at Hunsaker's insistence, a member of MIT's aeronautical engineering faculty, Hsue-shen Tsien, had begun preliminary studies for a hypersonic wind tunnel to investigate aerodynamics at speeds above Mach 5. Tsien had received a master's degree in aeronautical engineering at MIT in 1936, gone on to Caltech for his PH.D., and returned to the MIT faculty in 1946. Hunsaker suggested to George Lewis in May 1947 that MIT and the NACA cooperate in a joint research effort, with Tsien going to Langley to talk to people also interested in hypersonics. Hunsaker had two motives: A cooperative program would yield important research data that could be useful for future high-speed aircraft and missiles, but more important, it would help stave off what appeared to him to be a political move by the AAF to steal the march on the NACA in a potentially vital area of basic research.[28]

Despite Hunsaker's desire to move quickly with the unitary plan, the Truman administration decided to wait until it had a chance to hear from industry and the military. The President's Air Policy Commission, headed by Thomas K.

Finletter, convened during the fall of 1947. Deliberating at a time of crisis, with relations between the United States and the Soviet Union rapidly deteriorating into the cold war, the Finletter Commission explored all aspects of aviation and national defense. Hunsaker met with Finletter in September to discuss the problems of the aircraft industry and the roles of the NACA and the military in aeronautical research, stressing the need to put the unitary plan into effect soon. In its 1 January 1948 report (with the apocalyptic title, *Survival in the Air Age*), the Finletter Commission emphasized the military's responsibilities for aeronautical research and identified nuclear propulsion, electronics, and guided weapons as areas needing the most attention. The NACA's job, according to the commission, was to do the basic research in these fields, providing it to the military and industry in a timely manner so as to support their developmental work. The report also called for increased appropriations for aeronautical research and development.[29]

Laid up in a Brookline hospital recuperating from gall bladder surgery in February 1948, Hunsaker was not able to respond as soon as he would have liked to the Finletter report. "All repaired now and good for another 100,000 miles," he was back to work in March hoping he could use the commission's findings as leverage to get more money for the NACA and to move the unitary plan closer to reality. In a letter to one of the assistant directors of the Bureau of the Budget, he said that the NACA was "gratified" that the commission wanted more money to be allocated for aeronautical research and pointed out that the NACA was already vigorously pursuing research in the fields identified as top priority by the panel. He urged the bureau to act quickly on the NACA's recommendations for additional money for research facilities at universities and at NACA labs.[30]

In 1949 a facilities panel headed by James H. ("Jimmy") Doolittle made final decisions about who got what wind tunnels when. A pioneer aviator holding a doctorate from MIT, hero of the 1942 Tokyo Raid, commanding officer of the Eighth Air Force in England during the war, and a vice president and director of the Shell Oil Company, Doolittle had come onto the NACA Main Committee in 1948. Hunsaker considered him one of his best friends and enjoyed his companionship on fishing trips to Maine and other places. After discussions with Hunsaker, Dryden, and members of the NACA Executive Committee, the Doolittle panel's recommendations were integrated into the estimates forwarded to the Bureau of the Budget for fiscal year 1951. As revised for what Hunsaker hoped was the last time, the plan included thirteen tunnels for universities, twelve of them small tunnels following standardized blueprints developed at the NACA Langley lab to save money, and the remaining one to be a larger and more expensive hypersonic tunnel. At long last, just before the end of the legislative session in 1949, Congress took action on the NACA's recommendations, voting in October to approve a much-amended version of the unitary plan bill.[31]

The total initial NACA allotment under the unitary program came to $136 million, specifically earmarked for an eleven-by-eleven-foot (3.4-by-3.4-meter) transonic tunnel and two supersonic tunnels (nine-by-seven feet and eight-by-seven feet [2.7-by-2.1 meters and 2.4-by-2.1 meters]) at Ames, a ten-by-ten-foot (3-by-3-meter) supersonic tunnel at Cleveland, and a four-by-four-foot (1.2-by-1.2-meter) supersonic tunnel at Langley. The navy also got a unitary tunnel at the David W. Taylor Model Basin. The air force, independent for the last two years within the new Department of Defense, secured an authorization for $100 million for the AEDC at Tullahoma, to be spent on the transonic propulsion tunnel (reduced in scale to sixteen-by-sixteen feet [4.9-by-4.9 meters]) and the controversial hypersonic tunnel (the same dimensions but with its speed cut to just below Mach 5).[32]

That the unitary plan turned out to be neither unitary nor much of a plan was not Hunsaker's or the NACA's fault. In part, the plan was a legacy of the rivalry between Hunsaker's old navy-MIT-NACA axis and that of von Kármán, the air force, and Caltech. Hunsaker saw the plan as the only viable response to what he considered a bid for empire by the AAF at the expense of the NACA. In another

Hunsaker chairing a meeting of the NACA Executive Committee, April 1950. (255-GF-409, Courtesy National Archives)

sense, the air force's aerodynamic research program looked like the ghost of Billy Mitchell coming back to haunt him. As a battle-scarred veteran of the struggles against a unified air force in 1919–25, Hunsaker was determined to block all such efforts at aggrandizement by the air power advocates in the postwar years.

Though a true believer in the power and potential of aviation, Hunsaker never reconciled himself to the air force's post-1945 commitment to strategic bombing with nuclear weapons. He was appalled in April 1949 when Secretary of Defense Louis Johnson canceled the navy's new supercarrier in favor of the air force's intercontinental-range B-36 nuclear bomber. Things got much worse later in the year with the "revolt of the admirals." Navy accusations of irregularities in the B-36 procurement led the House to hold hearings on the matter, subjecting the service to intense criticism from the air force, Congress, the Truman administration, and the press, and leading to the resignation of the chief of Naval Operations, Admiral Louis E. Denfeld.[33]

Hunsaker wrote to a friend in November 1949 that the B-36 was "the dream of the wild blue yonder boys, the refuge of the pacifists, one-worlders, isolationists, and jingos" who were blind to the country's need for strong conventional forces. Two years later he was no more convinced by the air force's "craze" for "pushbutton warfare"; he doubted that the B-36 could do what the air force claimed and questioned the Strategic Air Command's commitment to destroying the Soviet Union's war-fighting potential. Even if that were possible, he concluded that it was a "pretty drastic" way to win a war and probably would not have the full support of American public opinion. Hunsaker had learned that it was politically impossible to oppose the service's plans directly, especially in 1947 when it was a foregone conclusion that the army's air arm would become the independent air force.[34] In the chilly domestic and international climate of the cold war, it took someone with Hunsaker's formidable political abilities to prevent the air force from heading off on its own research program and securing its cooperation with the NACA, even though in the end the service got most of what it wanted at Tullahoma.

That Hunsaker was a strong and consistent voice in support of the navy and its air arm as a vital component of the nation's defense forces should surprise no one. The Korean War (1950–53) verified his ideas on the need for a balanced national defense. Only days after the start of the conflict, he placed the blame for American lack of preparedness on the air force's addiction to the strategic bombing concept, which had become a convenient excuse for not having "an all-round Army and Navy." His prediction that there would be a "reversal of policy regarding 'ground support' airplanes and weapons" was largely borne out as the war dragged on and the air force had to devise means of close air support. He remarked in 1954 that the air force was well prepared for intercontinental warfare, "if unopposed," and criticized the service for its "theory of a 30-day war,' won by it alone." To offset the air force's near monopoly on strategic warfare and to restore an equilibrium to the nation's force structure, the navy had developed the aircraft carrier's nuclear strike capability, improved fleet air defense, and introduced a promising new generation of land attack missiles.[35]

Hunsaker's chairmanship of the NACA and close contacts with the Atomic Energy Commission (AEC) took him into aircraft nuclear propulsion. In the awful aftermath of Hiroshima and Nagasaki many scientists and engineers saw nuclear fission as the technological Holy Grail, a boundless source of energy on earth as well as in the skies. A lightweight reactor generating heat to drive turbines and produce thrust could provide speed and endurance beyond that possible with conventional aircraft power plants. In his April 1946 National Academy of Sciences paper on propulsion, Hunsaker suggested that there might be ways of using atomic energy to accelerate electrons or gaseous ions, cautioning that the high temperatures involved would require major breakthroughs in materials technology. The following year, Hunsaker joined the AEC's Industrial Advisory Group, which debated how much authority should be centralized in the AEC and how much distributed among the various agencies involved in nuclear aircraft development work. It was agreed that the AEC should function as the coordinating agency, with key people assigned to the commission from the other organizations involved in aircraft nuclear propulsion projects.[36]

Exactly when the NACA became involved in the nuclear aircraft program is unclear. In September 1945, only weeks after the atomic fireballs ended the war in the Pacific, Hunsaker reported to the NACA Executive Committee that the British had suggested "that a study be made of the possibilities of atomic energy to determine whether there is any prospect of its useful adaptation for aircraft, apart from destruction." Well before the end of the war, key people at the Cleveland propulsion lab envisioned the potential of nuclear power for aircraft and suggested that the NACA work closely with its Manhattan Project counterparts. At Hunsaker's urging and with the support of the physicist Edward U. Condon, director of the National Bureau of Standards, the NACA in 1947 set up a broad-based nuclear power committee, consisting of representatives of the war department, the navy, the AEC, and universities, to investigate the problem. Finally, in July 1948, the NACA and the AEC reached an agreement to cooperate on the nuclear aircraft propulsion program.[37]

Through the remainder of the 1940s and into the 1950s, as the NACA nuclear program continued in cooperation with the AEC and the air force, Hunsaker remained committed to the concept despite its complexities and the monumental technical hurdles that had to be overcome. In the fall of 1954 he told AEC Chairman Lewis L. Strauss that he thought the NACA needed to proceed with a new component research facility for aircraft nuclear propulsion, in which the NACA would study the effects of radiation and high temperature on power plant materials. Located at the former Plum Brook Arsenal near Sandusky, Ohio, the $4.5 million NACA lab featured a research reactor and related equipment needed to explore the myriad technical problems associated with the aircraft nuclear propulsion program.[38]

Drawn to the vision of flight at virtually unlimited range and uneasy about the nation's defense, Hunsaker remained a firm supporter of the nuclear aircraft program despite its slow pace of development and high costs. In a 1955 history of the NACA, he acknowledged the problems associated with the development of aircraft nuclear propulsion but concluded that "our national security requires that the research and development of nuclear power plants for aircraft be carried forward with unceasing vigor." In January 1956, when he forwarded the NACA's annual report to Congress and the White House, he emphasized the importance of the nuclear-powered airplane to the national defense and warned that the Soviet Union might be leading the United States in this vital area of research.[39]

Hunsaker's uncritical and overly enthusiastic support of nuclear-powered aircraft, which persisted for some time after many other engineers had rightly concluded that it was technically and economically unfeasible, reveals another of his blind spots. He did not, however, continue his advocacy of the nuclear airplane with the single-minded purpose that he had advocated the rigid airship two decades earlier. After leaving the NACA chairmanship, he headed an ad hoc committee charged with examining the hazards of aircraft nuclear propulsion. Beginning in November 1957, the Hunsaker committee heard from representatives of the air force, navy, and private contractors about nearly overwhelming problems: base siting, operational difficulties, the need for thirty-minute alert readiness, and crew radiation safety. By far the worst was the ever-present danger of an aircraft crash, which would spread lethal doses of radiation over one square mile (2.6 square kilometers), require the cordoning off of another ten to twenty-five square miles (25.9 to 64.7 square kilometers), and expose people within a hundred square miles (259 square kilometers) to unacceptable radiation dosages. This was all Hunsaker needed to convince him and other members of the committee that the NACA and the armed forces were pursuing a dangerous and expensive chimera.[40]

During the 1940s and early 1950s Hunsaker also focused his attention on air transport safety, something that had long interested him and that took on more urgency as the expansion of commercial air traffic generated congestion and delays at the nation's major airports. Following a spate of fatal crashes in the fall and early winter of 1946, the Republican Congress launched a series of investigations, finding that "overregulation" by the CAA was largely to blame for the problems in air transport. Unwilling to let a hostile Congress steal the spotlight on commercial aviation safety, President Truman created the Presidential Board of Inquiry on Air Safety, chaired by James M. Landis, head of the Civil Aeronautics Board, on 15 June 1947. Hunsaker served as vice chairman of the board, which included among its other members CAA administrator Theodore P. Wright and representatives of the Air Transport Association and the Air Line Pilots Association. Presenting its findings on 29 December, the board concluded that most of the culpability for safety problems lay with the airlines, directly contradicting the conclusions of the congressional investigating committees. Much to Hunsaker's delight, the report urged the federal government to put more money into research by the CAA and the NACA.[41]

In and around Elizabeth, New Jersey, in late 1951 and early 1952 another terrible rash of fatal accidents brought air safety once more to the attention of the public and official Washington. As the death toll mounted in the air and on the ground, local authorities took the extreme measure of closing busy Newark Airport while the CAA looked at what it could do immediately to improve safety in the skies over metropolitan New York and to restore public confidence in the air transport system. As he had in 1947, Truman appointed a special committee to study the problem and make recommendations. Announced on 20 February 1952, the President's Airport Commission included Jimmy Doolittle as chair and Hunsaker and CAA administrator Charles F. Horne as the two members. With the public and the president clamoring for quick answers, the commission got to work within a week, bringing in advisers from the CAA and the NACA and setting up offices and a work schedule. For three weeks in March, the commission heard testimony from experts on safety and representatives from the industry, followed by visits to sixteen of the nation's busiest airports. Because of his familiarity with the area, Hunsaker inspected Boston's airport and talked with municipal officials there before joining Doolittle for a tour of civil and military airports in the Washington-Baltimore region.[42]

The commission's report, *The Airport and Its Neighbors,* issued on 16 May 1952, criticized the lack of coordination at the federal, state, and municipal levels in implementing the Federal Aid Airport Program and the failure to incorporate airport construction into local planning efforts. The report was also critical of Congress for not providing the money to do all that was needed to upgrade terminal and en route navigation, air traffic control, and safety systems. Among its recommendations, the commission called for the "highest priority" to be given to federal assistance for runway construction, new zoning laws and tougher enforcement to lessen the encroachment of residential and commercial buildings, and the integration of municipal and airport planning. Additionally, the report called for the development of new technology to effect positive air traffic control at each of the nation's major airports. The commission's hard work seemed to go for naught, however, because Truman sat on the report for weeks before forwarding it to the relevant agencies, which could do little to implement it as a consequence of decreased federal airport funding in 1953.[43]

In the years immediately following World War II, there was also much to occupy Hunsaker at MIT, which had consolidated its position as one of the nation's foremost research institutions. In November 1945 the navy contracted with MIT to develop ground-to-air and air-to-air missiles for the Bureau of Ordnance. Project Meteor, as it was known, was a multimillion-dollar cooperative effort involving the aeronautical, electrical, and chemical engineering departments in basic research as well as component design and vehicle flight testing. Through the first six months of the project, most of the emphasis was on aerodynamics, control, and propulsion, followed by work with two "associate contractors," United Aircraft and Bell Aircraft, both of which had considerable experience in high-speed flight. As the project matured, the air-to-air missile assumed priority over the ground-to-air weapon.[44]

From the start, Hunsaker realized that Meteor presented an opportunity for his department to obtain the aerodynamic research facilities it needed to study transonic and supersonic flight. The Bureau of Ordnance allocated $14 million for the program, of which $2.6 million was set aside for a supersonic wind tunnel. Following the design of one of the new supersonic tunnels at the NACA's Ames laboratory, the MIT facility was a 10,000-horsepower variable-density Mach 3 tunnel with a three-foot-square (0.91- meter-square) test section that could be reduced to two square feet (0.6 square meter) for higher speeds. Groundbreaking for the tunnel, located on the west side of campus, was on 14 June—alumni day at MIT—in 1947, with dedication as the Naval Supersonic Wind Tunnel on 1 December 1949. Even though Project Meteor ended in 1953 without producing an operational weapon, the wind tunnel was invaluable in attracting other high-speed research projects from industry and the government.[45]

At MIT Hunsaker became embroiled in the anticommunist hysteria that swept through the nation's campuses in the late 1940s and early 1950s. A Republican who came to adulthood in the era of Theodore Roosevelt, patriotic in an old-fashioned way, and committed to service at all levels of government, Hunsaker was a dedicated cold warrior. Nevertheless, like many in academe he found himself in an ambiguous position when congressional probes into alleged communist activities in the United States clashed with his professional and academic sensibilities.

Edward Condon was among those who came under scrutiny by the House Subcommittee on National Security, chaired by Republican J. Parnell Thomas of New Jersey. Alfred N. Richards, the president of the National Academy of Sciences, circulated a draft statement of protest in March 1948 among the members of the academy's Advisory Council, inviting their comments about its content and possible publication. Hunsaker recommended that the Advisory Council alone publish the statement and that it not come from the National Academy of Sciences as a whole; others on the Advisory Council opposed publishing any statement at all. Finally, Richards, concluding that circulating an unsolicited protest, however mild, would "jeopardize our relations with Government," decided that the statement should go privately to the Thomas committee. Hunsaker saw little purpose in bearding Thomas in his den and argued for presenting the statement in person to the Speaker of the House. In the end, the National Academy of Sciences did nothing.[46]

In public neither the National Academy of Sciences nor Hunsaker said or did anything to defend one of their own. But in private Hunsaker expressed strong views on the Condon case. At the end of March 1948 he wrote to Edward P. Farley, a friend from the Goodyear-Zeppelin days, that Condon was a "fine man," whose only offense had been to associate with people in the Polish and Soviet embassies during the war. "Some of these people may be bad medicine," Hunsaker conceded, "but if everyone who had a drink with them is to be smeared they would fire a Ballroom. I have known Condon well and favorably and this is a good case on which to register indignation at a type of Congressional witch hunting."[47]

Wisconsin Senator Joseph McCarthy's accusations of communist infiltration into the highest levels of government accelerated the witch-hunting Hunsaker deplored. At MIT Professor Dirk Jan Struik, an internationally recognized mathematician and avowed Stalinist, came under the scrutiny of the Massachusetts Special Commission on Communism and was indicted in September 1951 for conspiring to overthrow the government of the commonwealth. Hunsaker thought the allegations were absurd but still voted with other MIT faculty to suspend Struik with pay from his teaching responsibilities pending the final determination of the case. In 1952, hoping to end such ridiculous red-baiting and realizing that the Struik case had serious repercussions for academic freedom, Hunsaker sent money to the Struik Defense Committee. Eventually, MIT reinstated Struik, but the case dragged on until 1956, when it was rightfully dismissed by the United States Supreme Court.[48]

Travel abroad was a welcome break from Hunsaker's responsibilities at MIT and the NACA. In 1948 he received an invitation to attend the Congress of the International Union of Theoretical and Applied Mechanics at the Imperial College of Science and Technology in London. It took some schedule juggling before Hunsaker finally decided he could go, traveling with Alice and using the trip as an excuse to see "little" Alice and her family, which had expanded with the birth of a son. The Hunsakers again opted not to fly across the Atlantic, sailing on 21 August aboard the giant Cunarder *Queen Elizabeth* and arriving in London on the twenty-sixth. Hunsaker attended the conference from 5 to 11 September, and with Hugh Dryden made side trips to glean whatever new information he could from his British counterparts.[49]

Hunsaker was also part of a big MIT and NACA contingent at the International Union's Eighth International Congress in Istanbul in 1952. Again sailing on the *Queen Mary,* he combined the trip with a couple of weeks' vacation in France, taking Alice and her younger sister Melanie ("Mimi") Avery, who had been living with them in the downstairs apartment at 10 Louisburg Square. From France, the party flew to Istanbul for the start of the congress on 20 August. On the second day of the meeting, Hunsaker delivered a lecture on the social aspects of aeronautics. To him, modern air transportation allowed people to traverse the "ocean of air," to meet and mix with other people, helping to break down national "frontiers" that inhibited peace and understanding. In this context the Iron Curtain ran contrary to progress and modernity, amounting to "an artificial and socially expensive attempt to maintain a geographical status more appropriate to a previous century."[50]

Once Hunsaker's obligations were over at the congress, he, Alice, and Mimi joined others in the group for tours of Istanbul and an excursion across the Bosporus to the Bithynian Olympus, reputed to be the "original" home of the Greek gods. Leaving Turkey behind, the Hunsakers went on to Paris for five days and then to London, arriving there on 17 September. They spent much of their time with Alice and her family in Kensington, and Hunsaker made official visits to the National Physical Laboratory, Farnborough, and the Vickers factory before the trio returned home on 29 September.[51]

Overseas trips were enjoyable to Hunsaker, as was recreational reading, usually limited (his leisure time being so scarce) to literature related to his work. In 1949 he discovered the work of Nevil Shute, pseudonym for Nevil Shute Norway. Shute's novel, *No Highway,* explored the dimensions of faith and rationality, capturing the dilemma engineers faced with metal fatigue in high-performance aircraft. There was "no known way," Hunsaker warned members of the NACA Executive Committee, "of detecting incipient fatigue failure, as material in which a large part of the fatigue life has already run looks as good as ever" and could occur at any time under the right combination of temperatures and loads. He also voraciously consumed popular American histories, paperback detective novels, and such specialist hobbyist magazines as the *Journal of the Flyfishers Club*.[52]

When he was not at MIT, in meetings, or traveling, Hunsaker was in the trenches in Washington testifying before Congress about the NACA's activities and funding requirements. From long experience, he knew how politics on "the Hill" worked and early on did well in getting the money the NACA needed. Total appropriations for the NACA fell from nearly $41 million in fiscal year 1945 to $24 million in fiscal year 1946, recovering to $30.7 million in fiscal year 1947 and to $43.4 million in fiscal year 1948. He appeared before the Independent Offices Appropriations Subcommittee of the House Appropriations Committee on 7 January 1948 to present the NACA estimates for fiscal year 1949. He asked for an increase of nearly $4.5 million from the 1948 appropriation for salaries and expenses, emphasizing that the additional money was essential for maintaining the research needed for national security. There had been a technological revolution in warfare since the end of World War II. The atomic bomb, jet propulsion, supersonic flight, and guided missiles taken together meant that the country could no longer count "on geographical isolation, on conventional weapons, nor on our industrial potential" for defense.[53]

The chairman of the subcommittee, Richard B. Wigglesworth, a Republican from Hunsaker's home state of Massachusetts, was sympathetic to Hunsaker and the NACA, although he and others did express concern about the expansion of the NACA budget and wondered whether some of its work paralleled that of other research organizations. Hunsaker assured the congressmen that the NACA's committee organization kept the air force and navy as well as other government agencies fully apprised of its research and that he was "reasonably" certain there was no unnecessary duplication. Much of the new money was needed for staffing, which Hunsaker considered vital to the efficient use of the NACA laboratories and equipment, and for the big electric utility bills incurred by the operation of powerful new wind tunnels. Hunsaker got what he came for; Congress appropriated $48.6 million for fiscal year 1949, of which $5.1 million went to salaries and expenses.[54]

Things did not go so well the following year, when Hunsaker ran afoul of the new subcommittee chairman, Albert F. Thomas, a budget-cutting Democrat from Houston who for years had been a behind-the-scenes power broker in Washington. In late February 1949 Hunsaker appeared before Thomas's subcommittee to justify an estimate of $48.7 million for the NACA's general expenses in fiscal year 1950, plus an additional $15 million for the construction of research facilities. Thomas thought the NACA had lacked congressional oversight and was dubious about its ponderous organizational structure and lax administration. Hunsaker's comparison of the Main Committee to a board of directors of a large corporation and his explanation that the demands on the NACA flowed from industry and the military astonished Thomas, who concluded that the "Committee and its activities stand out as unique" in his experience.[55]

Thomas then demanded that Hunsaker explain and account for nearly every dollar that he and the NACA were asking for, questioning the process that had led to the addition of two new members to the NACA Main Committee and the establishment of the research stations at Wallops Island and Muroc. The normally imperturbable Hunsaker bristled at the implication that "we have done something illegal and wrong." Thomas moved on to suggest that the NACA might save money by using funds transferred from the military services for the purchase of research aircraft rather than buying its own. To Hunsaker doing so would lead to a "bad management situation," with "several cooks" sharing responsibility, and would not save the taxpayer a penny. Thomas remained adamant that the NACA was riddled with inefficiency and waste; he cut its appropriation for salaries and expenses to $43 million and trimmed $5 million from the construction and equipment request.[56]

Calmer than the 1949 hearings, the January 1950 hearings on the NACA fiscal year 1951 budget revealed that Thomas was no more willing to go along with the organization's requests for more money than he had been previously. To counter Thomas's demands for budget cutting, Hunsaker brought along the air force and navy members of the Main Committee—Maj. Gen. Donald L. Putt and Vice Adm. John Dale Price—hoping to give Thomas an idea of the collaborative nature of the NACA organization. The fiscal year 1951 estimates were $46.1 million for salaries and expenses and $16.5 million for additional research facilities, for a total of $62.6 million. Thomas saw no reason why the NACA needed more money now, in time of peace, than it had during the war. Hunsaker replied that from 1941 to 1945 the NACA had deferred many research projects and that new technological challenges demanded more research money, and he contradicted Thomas's assertion that the nation had entered a new era of "permanent peace." The result was a standoff. Congress slightly cut the 1951 request for salaries and expenses to $45.75 million and increased the construction and equipment funding to $17.3 million—not nearly enough to make up for the cuts from the previous year. At Thomas's insistence Congress specifically excluded the 1951 funding for the NACA Paris office.[57]

Even though the NACA had done reasonably well with its fiscal year 1951 requests, the fight with Thomas over the 1950 appropriation rankled. Hunsaker wired Massachusetts Senator Leverett Saltonstall in August 1950 to register his dismay. The cuts, he told the Republican senator, "will have a crippling effect on the aeronautical research program of the National Advisory Committee for Aeronautics and consequently will be reflected by deficiency in quality of new aircraft to be provided [the] armed services by multibillion procurement funds."[58]

As if his previous committee experiences were not distressing enough, the 1951 hearings were much worse. During the hearings, held in the last week of February, Hunsaker asked for $55 million for salaries and expenses and $13.8 million for construction for fiscal year 1952. He patiently built a case for what he considered only modest increases over what the NACA had asked for the previous fiscal year, stressing the need for scientific research, emphasizing the growing threat of the Soviet Union, and explaining how the NACA formulated its budget. Incredulous that the NACA's expenditures had gone up "about 1,300 or 1,400 percent" in the last decade, Thomas said that he knew of "no government agency that even approximates the [NACA's] pattern of growth or rapidity of growth" and suggested that it might be time to abolish the NACA and turn its responsibilities over to private industry. Shocked, Hunsaker responded that "in some industrial concerns, which I naturally know as an engineer, the efficiency is at an incredibly low level" and that "it is economy for the Government to make use of a Government laboratory where we can." Thomas then assaulted Hunsaker with questions about the NACA's sick leave policy, its subsidization of chairs at universities, its travel expenses, its long-distance phone calls, and its use of airplanes for staff transportation.[59]

Any civility between the two evaporated when Thomas accused the NACA of having "too much fat" in its budget. Hunsaker snapped back at his antagonist that "we have not put any fat in here deliberately that you might get satisfaction from cutting out" and sarcastically thanked Thomas "very much" for his help in closing the NACA European office. When Thomas tried to settle him down by telling him to "keep your shoes on, Doctor," Hunsaker said that was hard to do when he was being charged with malfeasance. If that was what he thought, Thomas said, then maybe he should remove his shoes. Now barely able to contain himself, Hunsaker shouted that he would take off his shoes and his coat, too, "if you like," to which Thomas replied, "Let your size be your guide, my friend." Maybe Hunsaker's unusual loss of temper paid off; the NACA wound up with a small increase: $50.6 million for salaries and expenses and $18.3 million for construction and equipment.[60]

Welcome as it was, the 1952 increase in construction funds fell below what Hunsaker believed was needed for the NACA to staff its new research facilities. In January 1952 he appeared before Thomas's committee again to explain the NACA's activities and to justify its $73.2 million budget request for fiscal year 1953. Just as Hunsaker began his prepared statement, Thomas interrupted to ask him why the NACA had allowed American airplanes to fall to a "poor second best" in comparison to Russian-built models in combat in Korea. Hunsaker's anger flared again. Congress, which had not provided enough money for personnel or research, shared responsibility for any shortcomings. He had been "bitterly assailed" by the Mead committee in 1946 for "intolerable lack of imagination and leadership" in failing to direct the NACA toward more research, and now Thomas was implying he had failed again because the NACA emphasis on scientific research had prevented it from giving the air force the aircraft it needed. Thomas seemed uninterested as Hunsaker repeated the reasons that the NACA needed to upgrade its wind tunnels and other research facilities and needed more money for personnel. Once more Thomas exacted his pound of flesh; Congress appropriated $66.3 million for fiscal year 1953, nearly $3 million less than 1952 and far below Hunsaker's original request.[61]

For years Hunsaker's packed schedule had him shuttling between Cambridge and Washington, dividing his time as chairman of the NACA, his work on special committees, and his responsibilities at MIT as a faculty member and head of the Aeronautical Engineering Department, which had been spun off from the Mechanical Engineering Department in 1939. By the early 1950s he realized that he finally had to reduce some of his activities. As much as he enjoyed the stimulation of the classroom and the interaction with his colleagues in the Guggenheim Building, it was time to hand over his teaching and administrative loads to others. In 1951 he relinquished his position as head of the department to Charles Stark Draper, who had been in charge of the department's instrumentation laboratory, and the following year he gave up his regular classroom duties. Still vigorous and in good health at age sixty-five, he expected his reduced commitments at MIT would give him more time for the NACA and other professional activities and for reflection on the implications of what had been accomplished during his long career in aeronautics.

11.　　　The Long Twilight

he Hotel Statler in Boston was abuzz on the evening of 24 May 1952 as MIT's Aeronautical Engineering Department honored Jerome Hunsaker with a dinner marking the end of his nearly two decades of service to the institute. The banquet gave Hunsaker's friends and colleagues a chance to reminisce, get in some good-natured jibes, and say their farewells in a formal setting. John Markham, a member of the department since the 1920s, was the master of ceremonies. He singled out Hunsaker's early contributions to aerodynamics, which had allowed researchers for the first time to apply theory to practical problems of aircraft stability and control. "His return to the Institute in the early thirties was providential," Markham continued. "Under his direction the course in Aeronautical Engineering was strengthened and extended. Facilities for research were tremendously improved. He encouraged the training of more aeronautical engineers and particularly in graduate study." Markham knew that Hunsaker's retirement was conditional: "To speak of retirement for our colleague and boss for the past twenty years is amusing," for he and everyone else in the room knew that the last thing Hunsaker planned to do was to quit working.[1]

Markham was right: that retirement for Hunsaker was a relative term. For years he remained busy at MIT and the NACA, serving as a consultant for various companies with aviation and technical interests, and working on one committee after another; he was not ready to spend all his time hunting and fishing or retreating to Wildwood, his place at St. Huberts in the Adirondacks. His health sound and a chronic "workaholic" used to long hours, he went on for many years pursuing what he liked best—work. He also discovered, for perhaps the first time in his life, that his family could provide unexpected pleasures. He traveled more frequently with Alice and spent more time with his children and grandchildren.

Retaining the title of *lecturer* for five years after stepping down from his teaching and administrative positions, Hunsaker remained close to MIT and the department. He kept his office in the Guggenheim Building, ran a weekly seminar, which featured notables in all aspects of aviation, and kept involved in the school during his early retirement years by using his many contacts in the business and professional world to raise funds for special programs and endowed professorships.[2]

For money Hunsaker turned first to the Guggenheims. He had hoped in his last years as department head to interest Harry Guggenheim in committing the Daniel Guggenheim Fund to pay for an endowed chair in aeronautical engineering at MIT. When Hunsaker finally got to meet Guggenheim in the summer of 1948, he found him less than enthusiastic about the idea. Guggenheim thought the fund's money should go into rocket research, which he considered a "new and relatively uncultivated field," and he was not receptive to Hunsaker's pleas that traditional aeronautical fields had lacked governmental support since the war and needed private assistance. Hunsaker tried again in 1950 and 1951 to interest MIT's long-time benefactor in supporting a professorship in aeronautical engineering, only to see Guggenheim pledge the money to Caltech and Princeton, which used the funds to start rocket and jet research institutes (Caltech's evolved into today's Jet Propulsion Laboratory). Losing such plums to rivals on the West Coast and in the Ivy League was a major blow to MIT and a personal defeat for Hunsaker. The losses did nothing to change his mind about the new technology, and he was more determined than ever to put money into aeronautics as a means of restoring his and the department's prestige.[3]

Hunsaker turned next to H. Nelson Slater, a wealthy textile magnate and direct descendant of the famous admiral Horatio Nelson, the hero of Trafalgar, and Samuel Slater, who in the late eighteenth century brought to Pawtucket, Rhode Island, from England the secrets of mechanical cotton spinning. A frequent airline passenger and reserve naval officer, Nelson Slater agreed with Hunsaker that the rocket programs had been diverting some of the "best brains" from the problems of air transport, viewed an MIT professorship as a way of stimulating interest among students in a still-exciting field, and pledged $400,000 to endow the position. The first Slater post went to R. H. Miller.[4]

Another distinguished chair at MIT was the visiting Hunsaker Professorship in Aeronautical Engineering, established in 1954. With Lester D. Gardner, founder of *Aviation* magazine, leading the way, Hunsaker approached those he knew in the aviation world to pledge a half million dollars to endow the professorship. Eccentric aviation pioneer and airplane designer Glenn Martin provided much of the initial endowment. When he learned of the new professorship, Edward Warner wrote that "it certainly is right that your unparalleled services should be celebrated, even while you continue to teach, guide, and foster promising youth yourself for many years to come." William R. Hawthorne of Cambridge University, an expert on aircraft propulsion systems, was the first to hold the post, in 1955.[5]

The Hunsakers themselves endowed the Richard Cockburn Maclaurin Professorship in Aeronautics and Astronautics at MIT in late 1965. Alice had for some time wanted to do something to honor the former president of MIT who had done so much for the school and whose family she remained close to for fifty years. After meeting with MIT's president, James R. Killian, who thought it was a "heartwarming tribute" to one of the institute's great men, Hunsaker transferred $100,000 in stocks from his growing portfolio to begin the endowment, although some years passed before there was enough income to fund the chair.[6]

A continuing interest of Hunsaker in his later years was the National Academy of Sciences. He served on the academy's Membership Committee, recommending new people who might contribute to the organization and its activities, and even though he had stepped down as the academy's treasurer in 1948 he still offered timely advice on the academy's finances. In 1959 he became one of the academy's vice presidents. The connection suited both parties nicely. Hunsaker stayed actively involved in an organization that fostered interest in science and technology, and the academy benefited from Hunsaker's many personal connections in the academic, government, and business worlds.[7]

Despite having retired from the navy as a reserve captain in 1947, Hunsaker stayed close to the service and still had many friends there in the 1950s. James H. Smith, the assistant secretary of navy for air, phoned him in February 1955 to ask him to go along on a ten-day inspection tour of naval aviation activities in Europe and the Mediterranean. Hunsaker's group departed from Washington's National Airport on 3 September, making a fuel stop at Argentia, Newfoundland, before continuing on to London. Oddly, this was the first time Hunsaker—veteran transatlantic traveler, designer of the first airplane to fly the Atlantic, and planner of transatlantic airship operations—flew across the ocean. The party met with admiralty officials, getting a briefing on British guided missile developments, and visited the Royal Aircraft Establishment at Farnborough. They flew on to Brussels, Paris, Bonn, and Frankfurt, where they met with American naval air attachés, foreign service representatives, and naval officers assigned to NATO. The long flight back to Washington on 13–14 September included a stopover for fuel at Lages, in the Azores.[8]

As chairman of the NACA, Hunsaker continued to face questions about economy and efficiency that seemed to him to jeopardize the organization's research mission. In 1953 the General Accounting Office reported that the NACA had shifted money from salaries and expenses to pay for new construction and recommended that the NACA pay closer attention to its internal finances. Over the next few years, Hunsaker and the NACA tried to accommodate congressional demands for fiscal restraint, hoping to gain a more sympathetic ear from those holding the purse strings. It was to no avail, as annual budget appropriations for fiscal year 1954 fell to $51 million in salaries and expenses and to $11.4 million for construction. There was a crisis atmosphere in the summer of 1954 as the NACA struggled to reorder its priorities to fund a new initiative in high-speed aerodynamics. Dejected, Hunsaker wrote to Senator Leverett Saltonstall, "I evidently failed at the House hearing to make clear the drastic effect of withholding funds for staff to operate the new facilities to be completed in 1955." Realizing that there was only so much that could be done by the internal reallocation of funds, the Main Committee directed Hunsaker to take the NACA's request to President Dwight D. Eisenhower.[9]

Hunsaker wrote to the president on 6 August, telling him that recent aeronautical developments in the Soviet Union made it imperative that the NACA receive "additional funds . . . to operate our existing research facilities more intensively and also to provide for the construction of needed research facilities." Meeting with the president during the second week of August, Hunsaker and NACA research director Hugh Dryden made a convincing argument that the NACA urgently needed more money to deal with present and future problems in structures, high-speed aerodynamics, missiles, and other fields. Although Hunsaker's political skills had eluded him in debates with Thomas, they stood him in good stead with Eisenhower, who had recently set up a technological capabilities panel headed by James Killian and was increasingly worried about Soviet advances in long-range bombers and ballistic missiles. With Eisenhower's support, the NACA received a supplemental appropriation for its fiscal year 1955 budget, followed by a boost in its 1956 salaries and expenses to $60.1 million and construction and equipment to $12.6 million. These were the first increases in NACA appropriations since 1953.[10]

Chairmanship of the NACA and Hunsaker's many connections with official Washington led to other governmental responsibilities. Through long-time friend Luis de Florez, the brilliant, iconoclastic naval officer best known for Project Whirlwind, a general-purpose electronic digital computer first pro-

Hunsaker speaking at the formal dinner marking the fortieth anniversary of the NACA, 1955. Vannevar Bush, former chairman of the NACA, is on the right. The prestigious Collier Trophy is in the background. (255-GF-404, Courtesy National Archives)

posed for flight training simulators, Hunsaker became a member of a special science advisory board of the Central Intelligence Agency. De Florez's committee met three or four times annually in Washington to discuss how new technologies could assist CIA missions. Hunsaker recalled that he and the board could be "pretty inventive" in meeting the CIA's "almost impossible requirements" for operations within the Soviet bloc. Discussed by the panel were listening apparatus, methods of counting railroad cars and trucks entering and exiting from secret manufacturing facilities, forged documents, air insertion and recovery of agents and equipment, and the use of drugs "for specific effects." Another advisory committee for the CIA that Hunsaker served on in the 1950s was chaired by Cornelius Van S. Roosevelt, although there is nothing in the public record to indicate what that committee's activities were.[11]

Strategic reconnaissance was an important activity that overlapped Hunsaker's advisory obligations for the CIA and his position as NACA chairman. In 1954 the CIA determined that a high- altitude aircraft under development by Lockheed's "Skunk Works" might meet the country's demands for intelligence on Soviet activities in nuclear and long-range weapons systems. The result was the famous Lockheed U-2, an extraordinary airplane that flew at a height exceeding 70,000 feet (21,300 meters), well above Soviet antiaircraft defenses. Hunsaker learned of the U-2 in May 1956 and asked Hugh Dryden if he could see it when he went out to California for a meeting of the Institute of the Aeronautical Sciences. Dryden arranged for Hunsaker to visit Groom Lake, the Nevada test site where CIA pilots were undergoing the last stages of training before the airplane became operational. Hunsaker later agreed to let the NACA become part of the cover story for the U-2's clandestine overflights. As far as the world was concerned, the U-2 detachment that began flying out of England in June was participating in an NACA high-altitude meteorological research program.[12]

Hunsaker's name has been linked to another shadowy group, variously named Majestic 12, Majic 12, or MJ-12, which has recently attracted the attention of those determined to uncover what they fantasize as a vast multinational conspiracy to conceal the presence of alien spacecraft. These advocates of UFOS (unidentified flying objects) allege that Majestic 12 was formed in 1947 to collect and analyze alien technology and, through reverse engineering, to adapt it to military requirements. Hunsaker's connections to the military, scientific, and corporate communities made his name a natural for the committee. Because service on the supersecret Majestic 12 was supposedly a lifetime commitment, and because Hunsaker was the longest-surviving member, no documentary evidence of the organization surfaced until after his death in 1984. Doubters have found the provenance of Majestic 12 documents unclear at best, while supporters decry the skeptics' findings as further evidence of a massive official whitewash. Suffice it to say that in thousands of documents relating to Hunsaker in numerous archives there is not one reference or even allusion to such an organization or anything remotely like it. It is probable that Majestic 12 is nothing more than a figment of the imagination. It strains credulity that Hunsaker, whose life centered on reason and order, would ever assume that unusual aerial phenomena could only be the result of visits from outer space.[13]

With the NACA on the right track, at least as far as increased funding was concerned, Hunsaker in 1956 considered whether it was time to step down as NACA chairman. He had been thinking of resigning for some time and had various reasons for doing so, not the least of which was his advancing age. Still remarkably energetic for a man who had recently turned seventy, he nevertheless found it harder to keep up the pace of work he was used to. More important, he discerned a growing emphasis in the NACA on rockets and space. He was not opposed to research in this area, but he did not want international or domestic political circumstances to force the NACA to move in the direction of space or to otherwise dictate its research agenda. He told John Lansing Callan, a retired admiral and an old Curtiss aviator, that he remained "active enough" but did not want to stay on "when everyone is shouting about space missiles." Later he recalled that "space and rocket excitement was so hot" that it would likely obscure what he considered the more vital research in aerodynamics. Less charitably, he wrote to a friend in 1958 that "Space Cadet publicity," coupled with the influence of a group of "politically sensitive civilian leaders" and "young (and irresponsible) 'scientists'" were taking the old NACA in the wrong direction.[14]

Moreover, it was apparent to him that there was an emerging consensus that the committee form of organization was in all likelihood going to be cast aside in favor of a hierarchical agency headed by an administrator responsible not to a committee but to the president. Hunsaker viewed the NACA's governance by committee not as a weakness but as a strength. He had fiercely defended the autonomy and the organization of the NACA, fearing that its research agenda would be lost amid a welter of conflicting military and civilian priorities. In particular, he had used the power of consensus to negotiate a middle course between those who wanted the NACA to provide direct support to industry and the military in development programs and his own conviction that it should do all it could to nurture basic research. Rather than take the NACA through changes he thought were unwise, Hunsaker made his decision. At the 28 September 1956 meeting of the Executive Committee, just a month past his seventieth birthday, he announced his retirement. On 17 October, at the next Executive Committee meeting, Hunsaker handed over the NACA chairmanship to his friend Jimmy Doolittle.[15]

Over the next two years, Hunsaker saw the NACA head in exactly the direction he feared it would. The Soviet launch of Sputnik in October 1957 forced the Eisenhower administration to face the realities of the space age and raised the stock of the Young Turks, a new generation of NACA engineers and scientists who saw in astronautics the same excitement and promise that Hunsaker had discovered in aeronautics a half-century earlier. Staying on as a member of the Executive Committee, even chairing meetings in Doolittle's absence, Hunsaker still had a voice in the NACA. In 1958 the President's Science Advisory Committee recommended that a new executive agency be created, and a bill to establish the National Aeronautics and Space Administration (NASA) began to make its way through Congress. Hunsaker had deep reservations about the measure, calling it "a step backward," and said that he "spoke my piece with indignation" in a meeting with Killian. He believed the new agency would be prone to political interference and that it would be difficult to get bright, independent-minded engineers and scientists to serve on the proposed NASA Advisory Committee. At the 19 June 1958 Executive Committee meeting, Doolittle let his friend have his say, then told him that "the time has come for members of NACA to submerge their opinions and to make the proposed system operate as well as it possibly can." Hunsaker attended the next NACA meeting at Ames on 14 July but was absent from the 21 August meeting, chaired by T. Keith Glennan, NASA's first administrator.[16]

So it was that the NACA passed into history. The historian Alex Roland implies that Hunsaker either through naivete or self-deception compromised the NACA's independence, which he so cherished, and relegated the organization to subserviency under the military and industry. The result was that the NACA lost control of its research agenda and devolved into a "service agency spending most of its time on problem solving" for its "clients" in the military and industry. Roland adds that Hunsaker's personal conflicts of interest contributed to eroding the crucial autonomy of the NACA.[17]

To be sure, in the ethical climate of the early twenty-first century, Hunsaker could not have retained his business (and possibly even university) connections and still have assumed a position with a federal agency like the NACA. There were few scientists or engineers who knew the Washington merry-go-round as well as Hunsaker, who could be the master of expediency and was as comfortable among politicians and businessmen as he was in the world of academe. At the same time, his dedication to the public interest was unparalleled. He drove the NACA back in the direction of basic research in the immediate postwar years and steadfastly resisted industry representation on the Main Committee, while using his contacts and influence to satisfy the aircraft manufacturers' persistent demands for applied research. Nor is it fair to conclude that short-term compromises made by Hunsaker under wartime exigencies or the pressure of budget cuts meant that he abandoned the NACA's central research mission. More than anyone else, he understood the public responsibilities of the organization and stressed the long-term significance of scientific research.

Others go farther in their in criticism, maintaining that a symbiosis among scientists and engineers, capitalists, and federal bureaucrats, combined with perceived threats of communist subversion at home and Soviet expansionism abroad, led to a cynical alliance between capitalism and the military that endangered a free and democratic society. Moreover, they were compelled, as the historian David F. Noble argues, to defend their connections with political and business authority by appealing to progress as a higher moral and social goal than the sheer avaricious aggregation of power.[18]

There is nothing to suggest that Hunsaker's ties to the military-industrial-university complex threatened the independence of the NACA, degraded its position as one of the nation's most respected research organizations, or compromised American democracy. That Hunsaker was part of the complex, that he profited from it professionally and financially, and that he was a confirmed cold warrior who saw the Soviet Union as a threat is undeniable. But to conclude that he and others by design undermined the public interest in the quest for corporate benefit is an altogether different matter. Unless one assumes either that Hunsaker was deceitful or that he was somehow misled by his corporate masters, it is a safer bet that his work represented a sincere commitment to the nation and its people. Noble also fails to consider that scientists and engineers, whether in public service or not, go about their work out of dedication and because they like what they do. This was particularly true of Hunsaker, who like others in aviation looked beyond personal gain and viewed flight as intrinsically exciting and emotionally and intellectually rewarding.[19]

Certainly no revolutionary transformation of the NACA occurred during Hunsaker's watch; nor did he think such change was possible or necessary. Epitomizing the heterogeneous engineer, Hunsaker coupled his technological and organizational skills with a remarkable political instinct to help ensure that the NACA had enough money and facilities to retain its leadership in aerodynamics, advanced propulsion systems, and other areas he considered essential to the nation in times of continuing international crises. He was a visionary, paradoxically rooted in traditions dating from earlier in the century that did not always mesh well with the circumstances after 1950. Part of that vision was the realization that if it was time for the NACA to move on then others were better suited than he to take his old and familiar organization into new and mostly uncharted territory.

Richly deserved were the many honors and awards Hunsaker received in the 1950s as the aeronautical world bestowed on him its recognition for a lifetime of achievement. In 1951 he received one of the most prestigious prizes in aeronautics, the National Aeronautic Association's Wright Brothers Memorial Trophy "for significant public service of enduring value to aviation in the United States." In his acceptance speech during the banquet at Washington's Statler Hotel on 17 December, Hunsaker pointed out that the "convergence of two astounding revelations of technology," the airplane and the atomic bomb, had drastically altered traditional concepts of national security. No longer could the United States rely on geography and its enormous productive capacity to counter its cold war enemy, the Soviet Union. "We cannot afford to be second best in the air," he stressed, and through "intensive and imaginative research . . . we can and will keep our strong lead in the quality of American aircraft." The speech was a perceptive assessment of where the nation stood and the centrality of fundamental research in meeting the challenges of the cold war.[20]

Other accolades included the Godfrey L. Cabot Trophy from the Aero Club of New England as Aviation Man of the Year in 1953 and the Langley Gold Medal from the Smithsonian Institution in 1955—an honor bestowed on only a handful of luminaries in the aeronautical world. At a dinner and reception in the Great Hall of the Smithsonian "Castle" on the Mall on 14 April, Hunsaker graciously accepted the Langley medal, in recognition of his "superlatively important contributions to aeronautics" and his "scientific genius." From the NACA Hunsaker received the Distinguished Service Medal on 17 January 1957. The highest award given by the NACA, the medal noted Hunsaker's "service of fundamental significance to aeronautical science" and singled out his

> outstanding and unparalleled record of leadership during the past fifteen years as Chairman of the National Advisory Committee for Aeronautics. His brilliant contributions during World War II and the years of national emergency which lead [sic] to the era of supersonic flight disclosed on his part vision, judgment, and leadership of a high caliber, and a profound devotion to the public interest in the advancement of aeronautical science.

In May of the same year the Royal Aeronautical Society conferred on Hunsaker its Gold Medal for "his inspired chairmanship of the National Advisory Committee for Aeronautics, whose work has so greatly benefited aeronautical activities everywhere."[21]

Hunsaker had always had a feel for history, exemplified by his superb 1923 overview of naval aviation, but not until the 1950s did he have time to rediscover the past and contemplate how it informed the present. Hunsaker's 1952 book, *Aeronautics at the Mid-Century,* provided an opportunity for reflection and analysis of the airplane and aviation. An outgrowth of three lectures he had given in December 1951 at Yale University, the book was typical of Hunsaker's written work—well researched and precisely written. He emphasized how engineering science had advanced aerodynamics, propulsion, and control and guidance systems over the previous fifty years. He also looked at the evolution of air travel and the economics of air transportation, stressing the importance of government regulation and how such technologies as advanced autopilots, instrument landing systems, and radio navigation equipment had made air travel safer, cheaper, and more convenient. As for the military application of aviation, Hunsaker thought a small group of "enthusiasts" who had tried to "scuttle" the navy were now advocating almost sole reliance on strategic air power. "Such foolishness seems to erupt whenever new possibilities capture the imagination of undisciplined minds," he wrote. Finally, he cautioned planners against the panacea of high-technology air weapons and urged them to explore

Hunsaker receiving the Distinguished Service Medal, the highest award of the NACA, from his friend and successor as NACA chairman, Jimmy Doolittle, 1957. (255-GF-179-1, Courtesy National Archives)

ways of improving more traditional technologies and integrating them into the nation's defense.[22]

Aeronautics at the Mid-Century demonstrated that like many others of his generation Hunsaker viewed change through the unfiltered lens of progress. There was, he wrote, an innate "propelling force in technology" that led to economic and social betterment. He had no doubt that a linear progression toward faster and more efficient aircraft would generate economic prosperity and that "aeronautical research and engineering development will continue both to extend our knowledge and to apply it in new ways." As air transportation became more commonplace, he envisioned a process of homogenization, with "urban tastes" dominating human activities. New and larger airports would have major local and regional economic effects and change the spatial patterns in and around the cities they served. "With the national and international mobility afforded by air transport," he asserted, "national frontiers should be less marked by regional prejudices and less of a barrier to mutual understanding."[23] Hunsaker's unshakeable faith in progress was a sign of the times; most thoughtful Americans in the 1950s believed that advanced technology was one of humanity's best indicators of economic and social advancement. Nothing in Hunsaker's private or public writings indicated any critical reassessment of the broader meaning of technological change or the centrality of engineers in bringing it about.

Membership in the Society for the History of Technology also highlighted Hunsaker's growing interest in the social and economic effects of aviation and the wider significance of technology in general. In the early 1960s Melvin Kranzberg of Case Institute of Technology, who edited the society's journal *Technology and Culture,* persuaded Hunsaker to review books on the transatlantic flight of the NC-4 and a biography of Sir George Cayley, the early-nineteenth-century pioneer of aeronautical engineering. He also refereed at least one manuscript article and offered his advice on the appearance of the journal and the types of articles that it should run. A member of its advisory council since January 1961, Hunsaker's affiliation with the society was a productive, if brief, collaborative and interdisciplinary experience that widened the horizons of both parties.[24]

As the years slid by, Hunsaker found himself in demand as a resource person by researchers who wanted firsthand impressions of key people and events in aeronautical history. Lee Pearson, a historian with the Bureau of Aeronautics, recognized that Hunsaker by the late 1950s was one of the most valuable people left from the early years of naval aviation. In 1958 he met and discussed with Hunsaker technical developments in early naval aviation and the eventual disposition of his voluminous personal and professional papers. During later meetings between the two, they discussed specifics of aircraft design and changes in the administration of naval aviation. Pearson surveyed Hunsaker's papers and strongly advised him that he ought to make provisions for preserving them in a public repository for use by future researchers. He also cautioned Hunsaker not to segregate the personal information in the files or to distribute it among family members, for fear that it might be lost or destroyed. In the end, large portions of the files were deposited, per Hunsaker's wishes, to MIT and the Smithsonian's National Air and Space Museum.[25]

Another opportunity for reflection came when the Wings Club of New York City invited Hunsaker to deliver its second annual Sight Lecture in May 1965. The idea behind the lectures was to have participants look back at past accomplishments in aviation and forward to what they expected in the future. Now aged seventy-eight, Hunsaker eagerly grabbed the chance to survey his own career in aeronautics and offer thoughts about what lay ahead. The result was not only an unusual glimpse of the depth of his intellect but also of his ability to grasp the nuances of technology and policy. He summarized the European influence on American aerodynamics, the emergence of naval aviation in World

War I, the transatlantic flight of the NC-4, and advances in aircraft engines and fuels before turning to his involvement in lighter-than-air technology. With a tinge of remorse, he admitted that the rigid airship was a good example of "clear Hind Sight giving a very wrong Fore Sight"; in other words, the technology gave every appearance in the 1920s of solving the pressing needs of military and commercial aviation for more payload and range, but the results turned out to be a costly disaster.[26]

Looking ahead, Hunsaker remained confident that technical experts, civilian boards, and a combination of university and government research facilities were ideal for near-term technological progress, provided people did not confuse what bureaucracies wanted now with what the nation needed for its long-term security and well-being. Depth of experience and knowledge of history were vital to provide the perspective needed for longer-range vision. Often key individuals displayed the foresight needed to effect change, as evidenced by the strong personal role the NACA's George Lewis and Hugh Dryden played in co-ordinating government and industry in aeronautical research. Hunsaker understood that computers were valuable for facilitating decision making, but he worried that most of what went into computers came from less-experienced, junior people in organizations, who might not be able to see the big picture—so crucial for decision making. Two competing trends had to be balanced. On the one hand there was government support for research, absolutely essential in the cold war but that stifled innovation by directing inquiry toward what was needed now. On the other hand there was individual creativity and imagination, which looked beyond the immediate and practical requirements to developments that may not have any near-term application but that were vital to the country's future. "Let us never think we have no requirement for men with new ideas," he concluded.[27]

Intellectual reflections met the realities of the corporate world during the 1950s and 1960s through Hunsaker's continuing association with some of the nation's biggest companies. The Chrysler Corporation interested him more than others, largely because, among Detroit's "Big Three" automakers, the company had the best reputation for innovative engineering and because in the late 1950s its Defense Group was exploring ways of entering the promising areas of rockets and space. Hunsaker had worked as a consultant to Chrysler in 1940–41, had experience in high-speed aerodynamics, and was familiar with the missile guidance work of Charles Stark Draper, who had ties with Chrysler through the company's military contracts. Despite his skepticism about rockets and space, it did not take much arm-twisting for Hunsaker to be persuaded to join Chrysler Defense Group's Scientific Advisory Council in August 1958.[28]

Hunsaker at his desk at MIT, 1964. He lived for another twenty years and remained active for most of them. (JCH 13b, Courtesy MIT Museum)

The council explored a wide variety of topics at its two-day quarterly meetings over the next few years. Discussions centered on the relationship between automotive research and military power plants, gas turbines, Chrysler's Redstone and Jupiter ballistic missile projects, rocket staging, mobile missiles, inertial guidance systems, storable propellants, and the problems of controlling solid-fuel rockets. Hunsaker found the meetings "stimulating and of real interest," but he wondered what immediate practical value Chrysler's engineers got from them. It was important for the company to stay in NASA's good graces, while understanding that it was not likely to make a lot of money from NASA projects; he cautioned that the company should bid carefully on NASA contracts. On technical matters, he thought that liquid hydrogen fuel offered advantages in big space boosters and liked Chrysler's idea of clustering smaller rockets to achieve greater power rather than trying to develop larger rocket engines.[29]

Advancing age forced Hunsaker to cut back his many activities in the 1960s. One by one, he severed his corporate and institutional associations, starting in 1961 when he left the board of Tracerlab, a Boston research firm, and Goodyear, where he had been a director since 1946. The following year, he left the Chrysler Scientific Advisory Council and the consulting boards of Shell Oil Company and Sperry Rand, both of which he had joined in 1946, although he stayed on until the late 1960s as a director of both firms. He remained a director of McGraw-Hill for a few more years, counseling the big publishing house on technical and scientific works. He also continued as a regent of the Smithsonian Institution and vice president of the American Philosophical Society, associations he found consonant with his sense of the past while simultaneously keeping him in touch with the most recent trends in science and technology. As a trustee of the Boston Museum of Science in the 1950s, he helped with fundraising for the new Hayden Planetarium and after his resignation as a trustee in 1963 remained one of the museum's most generous contributors.[30]

With fewer distractions in his professional life, Hunsaker drew closer to Alice in his later years. The two traveled more often than they had in the past. In the winter of 1955 the couple spent time in Charleston, South Carolina, visiting Ralph Hanson from Hunsaker's master's degree days at MIT. That summer, Alice went alone across the Atlantic to visit young Alice and the Bird family, while Hunsaker was in California on official business with the NACA. In February 1956 they went south again, to the Pinckney Island Plantation, near Savannah, Georgia, with the Doolittles and other friends to hunt turkey and quail. At home in Louisburg Square, Alice kept up her interest in horticulture, even cajoling her husband into taking detours on auto trips to visit local gardens.[31]

Hunsaker's eclectic interests expanded to include environmental issues in the 1950s. An inveterate sportsman, he had always appreciated the wilderness and knew of the dangers to it from encroaching civilization. He also developed over the years a deep knowledge of North American flora and fauna. In June 1954 Hunsaker was on the West Coast for a meeting of the Institute of the Aeronautical Sciences at Los Angeles. After the meeting he and Robert C. Seamans Jr., a family friend, MIT graduate, and a systems engineer with Project Meteor, took off in a car to tour NACA and military flight testing facilities. Years later Seamans recalled with amazement how Hunsaker casually elaborated on the characteristics of various obscure species of cacti as they drove through California's high desert. Arranging a session on the "sea frontier" for the annual meeting of the American Association for the Advancement of Science in December 1953 brought Hunsaker into contact with Rachel Carson, whose book *Silent Spring* a decade later helped launch the environmental movement. Carson had just published her first major work, *The Sea Around Us,* and was already well known in oceanographic circles. In the 1960s Hunsaker continued to criticize the American space program, questioning the wisdom of spending billions to fly to the moon when there were so many more pressing problems on earth.[32]

Tragedy in one form or another often visits families, and Hunsaker's was no exception. The twins, Jerry Jr. and Peter, went their separate ways after graduating from high school. Jerry Jr. ultimately became a well-to-do broker, twice married and with three children. After his stint in the navy during the war, Peter—married with two children—worked for the engineering firm of Jackson and Moreland before joining Arthur D. Little in Cambridge as a project director. On 14 August 1959 Peter was at the Tech-Weld Corporation testing a liquid oxygen tank for a guided missile under development by his firm. Without warning, an inspection cover on the vessel burst, killing him almost instantly. Hunsaker received the news of the accident with shock and dismay, taking solace in knowing that his son's death was quick and painless. The death of a child before the parent seems to defy reason and the natural order of things; as the months went by, Hunsaker reconciled Peter's death in terms that seemed to make more sense, telling a friend that he considered his son to have been a "casualty of the Cold War."[33]

It almost sounds like name-dropping, but Hunsaker in retirement cemented friendships with the rich and famous—more famous than rich. He knew Paul Dudley White, the internationally recognized heart specialist, who was a neighbor in the Beacon Hill district. He continued his friendship with Jimmy Doolittle, which went back to Doolittle's years as an aeronautical engineering student at MIT in the mid-1920s. The two corresponded regularly, and Doolittle tried not to miss any of Hunsaker's spring fishing trips to Maine or winter quail-hunting expeditions to South Carolina and Georgia. Donald Douglas remained a close friend. Hunsaker wrote to him before the fall 1958 meeting of the MIT Corporation to invite him to stay in the downstairs apartment at 10 Louisburg Square and to join him on a round of parties and a tour of MIT, especially featuring the "dusty site where you and I worried over measuring damping derivatives for a wind tunnel model of a Jenny."[34]

Ever the devoted only child, Hunsaker tried as often as he could to visit his now much-aged mother, Alma, still living in the North Jefferson Street house in Saginaw. His father had been dead since 1939, but his mother went on in reasonably good health for many years, usually taking seasonal trips to Winter Park, Florida. In the summer of 1954, however, she suffered what most likely was a stroke, never fully recovered, and over the next four years lost her speech, sight, and hearing. Hunsaker began to see her monthly, sometimes combining trips to Saginaw with meetings of Chrysler's Scientific Advisory Council in Detroit. He admitted to his uncle Fred Clarke in January 1958 that he wondered how much good those visits were, even though he was encouraged that she was in no pain and expressed no anxiety about her condition. Lingering between life and death for years, Alma finally succumbed in 1962 at the age of ninety-nine.[35]

Alice enjoyed generally good health until she suffered a perforated duodenal ulcer and had to have major surgery in early 1959. The cause of the bleeding was most likely the aspirin she had been taking for arthritis. After a slow convalescence and recovery, she seemed to do well for a number of years. Then in August 1966, at St. Huberts, she fell ill with another duodenal hemorrhage and had to be taken to the hospital at Plattsburgh. Although surgery stopped the bleeding, the massive blood loss led to other complications, and she died in the hospital on 11 September at the age of seventy-nine. There was a memorial service for her in King's Chapel in Boston, attended by family and friends, and her remains were interred in the Porter family plot in Farmington, Connecticut. That Alice did not hang on or suffer at the end was some consolation, but the loss struck Hunsaker deeply. The Hunsakers' fifty-five-year marriage was a lifelong union of two highly intelligent people with different interests, its staying power bolstered by mutual respect and admiration.[36]

Alice's death left a void that could never be filled. Alone in the big house on Louisburg Square, Hunsaker had help from a housekeeper, and Jerry Jr. kept track of his finances and medical needs. There were still fishing trips and Wildwood in the Adirondacks to keep him interested, as well as activities with the family. Into the late 1960s, Hunsaker's Quartet Club (named for its four owners) got together to fish at Grand Lake, near Princeton in Maine. Other regulars in the informal group were Paul Dudley White, Boston banker J. M. Barker, and old friends Jerry Land and Jimmy Doolittle. A party in June 1968 at Jerry Jr.'s farm in Lincoln to celebrate a granddaughter's high school graduation was memorable for the excellent wine and brandy from Jerry's ample cellar.[37]

Long life has its rewards, but a price to pay is the steady and accumulating passing of old friends. George Westervelt went first, in 1956, followed by Luis de Florez in 1962, and less than a year later in 1963 by Julius Furer. Hugh Dryden's death in December 1965 after a long battle with cancer was upsetting. Hunsaker and Robert Seamans praised Dryden for his scientific research and the leadership he had provided the NACA in a memorial in the *American Philosophical Society Year Book* in 1966. Jerry Land died in 1971, and Vannevar Bush in 1974.[38]

Hunsaker's own health began a gradual and inexorable decline in the late 1950s. He suffered from what he thought were angina attacks in the late 1950s and told a friend in 1958 that it was "a warning of some kind. . . . However, if I don't climb mountains or get in a fight it may keep quiet." For another half-decade or so he had no major physical problems. Then sometime in the early 1960s, according to his surviving children, he suffered a stroke, which though not debilitating, led to the loss of short-term memory and personality changes that were noticeable to his family if not to those associated with him professionally. In 1972 he had surgery to repair a broken hip. He said the operation had left him "lame" and not able to get around as much as he had been used to. More than half a year later, the hip was still giving him trouble and continued to bother him through the remainder of his life. Despite poor eyesight and mental deterioration, which often left him confused, he sometimes insisted on driving, frightening passengers with occasional excursions off the beaten path into tennis courts and the forest surrounding the summer place at St. Huberts.[39]

Finally, and perhaps mercifully, as his physical and mental debilities multiplied, Hunsaker's stout heart gave out and he died at home on 10 September 1984, age ninety-eight. Eight days later, following his cremation, a memorial service was held at the nondenominational chapel at MIT with the minister of King's Chapel presiding. Unsure what to say about the passing of a man nearly a century old who held no religious beliefs, the minister read a eulogy from Robert G. Ingersoll, a nineteenth-century politician and popular lecturer, which expressed Hunsaker's and other's doubts about life on earth and the eternity beyond. "Every cradle asks us 'whence?' and every coffin, 'whither?'. . . I had rather live and love where death is King than have eternal life where love is not." His ashes were buried next to those of Alice in Farmington. A handsome slate tablet with a flowing symbol of flight engraved on it memorialized his nearly lifelong dedication to solving the mysteries of the air. Robert Seamans wrote the memoir for the American Philosophical Society. He emphasized Hunsaker's many contributions to aeronautics and his long service to his country as a naval officer, pioneer aerodynamicist, educator, and public administrator.[40] Yet more important may have been the bonds of professional and personal friendship that Hunsaker fostered within the aeronautical community over more than six decades, carrying forth his memory as a testament to new generations dedicated to understanding and mastering the skies.

The legacy of a man with such a formidable intellect who lived so long and did so much is not easily measured. No more than history can recreate the past, biography—even at its best—cannot restore life to a human being. All biography can do is provide an opportunity to lend shape and substance to a person's existence and by so molding that life story use it as a vehicle for painting a broader historical picture. Hunsaker pointed the way to the future as a technologist with interests that spread far beyond his curiosity about what happened to an airplane in flight to encompass a wider world of science, public policy, education, and administration. His was a twentieth-century vision rooted in rationalism and order. That Hunsaker's view of the world was flawed or that he and others who shared his faith in reason and progress did not accomplish their goals seems less important in retrospect than their persistence in the effort.

The classic heterogeneous engineer, Hunsaker might also be characterized as a technological maestro, someone who organized and directed a symphony of technology and, transcending his own individual contributions to the profession, took pride in the orchestra's performance.[41] Jerome Hunsaker would have agreed. For much of his long life he sought to perfect a composition in aeronautics that, despite a discordant note or two, appealed to an audience far larger and more diverse than even he could have imagined when he first entered the world of the professional engineer.

NOTES

I. An Officer and an Engineer

1. Unidentified, undated clipping of obituary (probably the *Washington Post*), Jerome C. Hunsaker biographical file, National Air and Space Museum Library, Washington, D.C. (hereafter cited as NASM Library); Robert C. Seamans Jr., e-mail to author, 3 June 1999; Jerome C. Hunsaker, interview by Walter T. Bonney, 2 November 1971 (hereafter cited as Hunsaker interview by Bonney, 2 November 1971), pp. 12–13, Hunsaker, J. C., Biography File, History Office, National Aeronautics and Space Administration Headquarters, Washington, D.C. (hereafter cited as NASA History Office). For Taylor, see William Hovgaard, "Biographical Memoir of David Watson Taylor, 1864–1940," *National Academy of Sciences, Biographical Memoirs* (Washington, D.C.: National Academy of Sciences, 1943), 22:135–53.
2. Edwin T. Layton Jr., *The Revolt of the Engineers: Social Responsibility and the American Engineering Profession* (Cleveland: Press of Case Western Reserve University, 1971), vii–ix, 53–78, 134–53.
3. For aeronautics as high technology, see Robert C. Post, interview with Walter Vincenti, "What Engineers Know," *Invention and Technology* 12 (Winter 1997): 20–26; Mark Levinson, "Encounters between Engineering Science and American Engineering Education: The Postwar Textbook Revolution" (paper presented at the meeting of the Society for the History of Technology, Pasadena, Calif., 16–19 October 1997), 1.
4. JCH quotation, 15 November 1949, folder 34, box 1, Jerome C. Hunsaker Papers, MC 272, Massachusetts Institute of Technology, Institute Archives and Special Collections (hereafter cited as JCHP/MIT).

5. Donald MacKenzie, *Inventing Accuracy: A Historical Sociology of Nuclear Missile Guidance* (Cambridge: MIT Press, 1990), 28, 86–87, 189.

6. "Union County History," Union County (Iowa) Genealogical Society http://lserver.aea14.k12.ia.us/SWP/jbriley/ucgen/history.html; JCH, "Detroit, Creston, and Okoboji, 1898–1902," p. 2, folder 2, box 1, JCHP/MIT.

7. Alice Hunsaker Bird, *Hunsaker Family History* (Henley-on-Thames, England: Privately published, 1995), 1–15, 82–86.

8. Excerpts from *Who's Who in America, 1938–1939,* in series I, Biographical, folder 1, box 1, JCHP/MIT; JCH, "Detroit, Creston, and Okoboji, 1898–1902," pp. 1–2, folder 2, box 1, JCHP/MIT; Bird, *Hunsaker Family History,* 20–21.

9. "The Reminiscences of Jerome C. Hunsaker," interview by Donald Shaughnessy, April 1960, Aviation Project, Oral History Research Office, Columbia University, p. 1 (hereafter cited as Hunsaker interview, April 1960, OHRO, Columbia University); JCH, "Detroit, Creston, and Okoboji, 1898–1902," pp. 3–4, folder 2, box 1, JCHP/MIT; *Current Biography, 1942* (New York: H. W. Wilson, 1942), 401; Alice Hunsaker Bird to the author, 12 November 1996.

10. JCH, "Detroit, Creston, and Okoboji, 1898–1902," pp. 5–6, folder 2, box 1, JCHP/MIT.

11. Excerpts from *Who's Who in America, 1938–1939,* in series I, Biographical, folder 1, box 1, JCHP/MIT; James Cooke Mills, *History of Saginaw County, Michigan: Historical, Commercial, Biographical,* 2 vols. (Saginaw: Seeman and Peters, 1918), 1:652; Stuart D. Gross, *Saginaw: A History of the Land and the City* (Woodland Hills, Calif.: Windsor, 1980), 63–65; Jeremy W. Kilar, *Michigan's Lumbertowns: Lumbermen and Laborers in Saginaw, Bay City, and Muskegon, 1870–1905* (Detroit: Wayne State University Press, 1990), 272–73, 284–86.

12. Mills, *History of Saginaw County,* 1:652, 654–56; Kilar, *Michigan's Lumbertowns,* 275; Gross, *Saginaw,* 179; Hunsaker interview, April 1960, OHRO, Columbia University, p. 1; JCH, "Saginaw, Rest Island, Annapolis (1902–1904)," p. 1, folder 3, box 1, JCHP/MIT.

13. Clipping from *Saginaw News,* 12 September 1984, Hunsaker biographical file, NASM Library; Hunsaker interview, April 1960, OHRO, Columbia University, p. 1; Mills, *History of Saginaw County,* 293; JCH, "Saginaw, Rest Island, Annapolis (1902–1904)," pp. 1–5, folder 3, box 1, JCHP/MIT.

14. Hunsaker interview, April 1960, OHRO, Columbia University, p. 9; George Carroll Dyer, *The Amphibians Came to Conquer: The Story of Admiral Richmond Kelly Turner,* 2 vols. (Washington, D.C.: Department of the Navy, 1972), 1:9–11; clipping from *Saginaw News,* 12 September 1984, Hunsaker biographical file, NASM Library.

15. Hunsaker interview, April 1960, OHRO, Columbia University, p. 9; Jack Sweetman, *The U.S. Naval Academy: An Illustrated History* (Annapolis: Naval Institute Press, 1979), 142–44; JCH to Walter Hunsaker, 18 May 1904, folder 4, box 3; JCH, "Saginaw, Rest Island, Annapolis (1902–1904)," pp. 7–8, folder 3, box 1; both in JCHP/MIT; Record of Midshipman Jerome Clarke Hunsaker, RG 405,

William W. Jeffries Memorial Archives, U.S. Naval Academy (hereafter cited as Archives, USNA); Dyer, *Amphibians Came to Conquer,* 1:15.

16. JCH, "Saginaw, Rest Island, Annapolis (1902–1904)," pp. 7–8, folder 3, box 1, JCHP/MIT; Dyer, *Amphibians Came to Conquer,* 1:15.

17. *Annual Register of the United States Naval Academy, 1905–1906* (Washington, D.C.: Government Printing Office, 1905), 122; Conduct Roll, Record of Midshipman Jerome Clarke Hunsaker, RG 405, Archives, USNA; Hunsaker interview, April 1960, OHRO, Columbia University, pp. 1–2, 5; Jerome Clarke Hunsaker, Physical Record, 1904–1908, Record of Midshipman Jerome Clarke Hunsaker, RG 405, Archives, USNA; JCH, "Naval Academy, 1904–1908," pp. 1–2, folder 4, box 1, JCHP/MIT; *Lucky Bag* (Springfield, Mass.: F. A. Bassette, 1906), 13:225.

18. *Hartford* deck logs, 24 November 1904 to 9 September 1905, Record Group 24, National Archives (hereafter cited as RG 24, NA); Record of Proceedings of a Board of Medical Examiners convened at the Navy Yard, Boston, in the case of Midshipman Jerome C. Hunsaker, U.S. Navy, 23 March 1910, vol. 762, Hunsaker, Jerome, Proceedings of Naval and Marine Examining Boards, box 553, Records of the Office of the Judge Advocate General (Navy), Record Group 125, National Archives (hereafter cited as JAG Records (Navy), RG 125, NA); Hunsaker interview, April 1960, OHRO, Columbia University, pp. 3–4; Commanding Officer, Coast Squadron, North Atlantic Fleet to Superintendent, U.S. Naval Academy, 22 August 1905, Record of Midshipman Jerome Clarke Hunsaker, RG 405, Archives, USNA.

19. *Cleveland* deck logs, 31 May 1906 to 31 December 1906, RG 24, NA; JCH to Alma, 28 July 1906, folder 4, box 3, JCHP/MIT; Jerome Clarke Hunsaker, Record of Officers, U.S. Navy, file 6733, National Personnel Records Center, St. Louis, Mo. (hereafter cited as JCH Service Record, file 6733, NPRC); summer cruise, USS *Cleveland,* 1906, cruises aboard USS *California,* USS *Cleveland,* and USS *North Carolina,* 1906–9, folder 9, box 2, JCHP/MIT.

20. *Annual Register of the United States Naval Academy, 1906–1907* (Washington, D.C.: Government Printing Office, 1906), 133; *Annual Register of the United States Naval Academy, 1907–1908* (Washington, D.C.: Government Printing Office, 1907), 117; JCH to Walter Hunsaker, 16 January 1907, folder 4, box 3, JCHP/MIT; Conduct Roll; JCH to Commandant of Midshipmen, 15 June 1907; both in Record of Midshipman Jerome Clarke Hunsaker, RG 405, Archives, USNA; Hunsaker interview, April 1960, OHRO, Columbia University, p. 2.

21. *Arkansas* deck logs, 1 June 1907 to 31 December 1907, RG 24, NA; JCH Service Record, file 6733, NPRC; Paolo E. Coletta, *Admiral Bradley A. Fiske and the American Navy* (Lawrence: Regents Press of Kansas, 1979), 81; Bradley A. Fiske, "Navigating without Horizon," *United States Naval Institute Proceedings* 33 (September 1907): 955–57, signed copy in folder 4, box 3, JCHP/MIT.

22. *Lucky Bag* (Philadelphia: Hoskins Press, 1908), 15:85; *Annual Register of the United States Naval Academy, 1908–1909* (Washington, D.C.: Government Printing Office, 1908), 52–53.

23. William A. Smith to Walter J. Hunsaker, 30 June 1908; J. E. Pillsbury (Acting Sec-Nav) to Walter J. Hunsaker, 8 July 1908; George A. Loud to Walter J. Hunsaker, 20 July 1908; all in folder 5, box 1, JCHP/MIT.

24. Journal of Jerome C. Hunsaker, 31 August to 1 November 1908, cruises aboard USS *California,* USS *Cleveland,* and USS *North Carolina,* folder 9, box 2, JCHP/MIT; JCH, "Sea Duty, 1908–1909," pp. 2–4, folder 9, box 2, JCHP/MIT.

25. Journal of Jerome C. Hunsaker, 31 August to 1 November 1908, cruises aboard USS *California,* USS *Cleveland,* and USS *North Carolina,* folder 9, box 2, JCHP/MIT.

26. Newspaper clippings in Alice Hunsaker Bird, ed., *Jerome Clarke Hunsaker: A Life in Aeronautics* (Henley-on-Thames, England: Privately published, 1982), 7–10; journal of Jerome C. Hunsaker, 1 January to 30 April 1909, cruises aboard USS *California,* USS *Cleveland,* and USS *North Carolina,* folder 9, box 2, JCHP/MIT.

27. Quarterly Reports of Fitness of Midshipmen, 30 September 1908, 31 December 1908, 31 March 1909, Record of Midshipman Jerome Clarke Hunsaker, RG 405, Archives, USNA; Hunsaker interview, April 1960, OHRO, Columbia University, p. 9; JCH, "Sea Duty, 1908–1909," p. 5, folder 9, box 2, JCHP/MIT.

28. Maurice Holland, *Architects of Aviation* (New York: Duell, Sloan, and Pearce, 1951), 9; JCH, "How did JCH get transferred to Construction Corps," pp. 1–3, folder 5, box 1, JCHP/MIT; JCH Service Record, file 6733, NPRC.

29. Samuel C. Prescott, *When MIT Was "Boston Tech," 1861–1916* (Cambridge, Mass.: Technology Press, 1954), 96–97.

30. Ibid., 159–61. For a perspective on such cooperative programs, see W. Bernard Carlson, "Academic Entrepreneurship and Engineering Education: Dugald C. Jackson and MIT-GE Cooperative Engineering Course, 1907–1932," *Technology and Culture* 29 (July 1988): 536–67.

31. Prescott, *When MIT Was "Boston Tech,"* 160–61; Hunsaker interview, April 1960, OHRO, Columbia University, p. 11; "Distinguished Dekes of Today," *Delta Kappa Epsilon Quarterly* 53 (May 1935): 61, folder 1, box 1, JCHP/MIT; JCH Service Record, file 6733, NPRC.

32. JCH, "How did JCH get transferred to Construction Corps," pp. 2–3, folder 5, box 1, JCHP/MIT; Bird, *Hunsaker Family History,* 35–39, 44, 58–59, 62–63, 95; JCH, "Family Narrative, 1909–1916," p. 1, folder 5, box 1, JCHP/MIT; JCH, "Miss Porter's School, Farmington, Connecticut," p. 1, folder 5, box 1, JCHP/MIT.

33. JCH, "Family Narrative, 1909–1916," p. 2, folder 5, box 1, JCHP/MIT.

34. Robert Kanigel, *The One Best Way: Frederick Winslow Taylor and the Enigma of Efficiency* (New York: Viking, 1997), 17–18, 449–57, 504–9; Hugh G. J. Aitken, *Scientific Management in Action: Taylorism at Watertown Arsenal,* rev. ed. (Princeton: Princeton University Press, 1985).

35. JCH, "Scientific Management," no date, folder 5, box 1, JCHP/MIT.

36. Ibid.; Aitken, *Scientific Management in Action,* 15–38, 132–35.

37. R. T. Hanson and J. C. Hunsaker, "Rudder Trials, USS *Sterett,*" *Transactions: Society of Naval Architects and Marine Engineers* (New York: Society of Naval Architects and Marine Engineers, 1912), 20:301, 303, 310, 332; *Dictionary of American Naval Fighting Ships* (Washington, D.C.: Naval History Division, 1976), 6:618–19; Hunsaker interview, April 1960, OHRO, Columbia University, pp. 11–12.
38. Hanson and Hunsaker, "Rudder Trials, USS *Sterett,*" 333–34.

2. Aeronautics

1. Hunsaker interview, April 1960, OHRO, Columbia University, pp. 6, 12–14; see also Earl A. Thornton, "MIT, Jerome C. Hunsaker, and the Origins of Aeronautical Engineering," *American Aviation Historical Society Journal* 43 (Winter 1998): 309.
2. Prescott, *When MIT Was "Boston Tech,"* 230–39; Hunsaker interview, April 1960, OHRO, Columbia University, pp. 16–17; Henry Greenleaf Pearson, *Richard Cockburn Maclaurin: President of the Massachusetts Institute of Technology* (New York: Macmillan, 1937), 51–56.
3. Hunsaker interview, April 1960, OHRO, Columbia University, p. 17; Shatswell Ober, "The Story of Aeronautics at MIT, 1895 to 1960" (Cambridge: MIT, Department of Aeronautics and Astronautics, 1965), 2–5; *Tech* 58 (11 March 1938): 1, 4; *Tech* 29 (14 October 1909): 1; *Tech* 29 (15 October 1909): 1; *Tech* 29 (4 April 1910): 1.
4. Pearson, *Richard Cockburn Maclaurin,* 242–43; Prescott, *When MIT Was "Boston Tech,"* 283; *Aircraft* 3 (September 1912): 225; Ober, "Story of Aeronautics at MIT," 4; Thornton, "MIT, Jerome C. Hunsaker, and the Origins of Aeronautical Engineering," 308.
5. E. Lapointe, "Aviation in the Navy," *United States Naval Institute Proceedings* 38 (June 1912): 627–56; Thornton, "MIT, Jerome C. Hunsaker, and the Origins of Aeronautical Engineering," 309.
6. John D. Anderson Jr., *A History of Aerodynamics and its Impact on Flying Machines* (Cambridge: Cambridge University Press, 1997), 268; JCH, handwritten response to Joseph S. Ames to JCH, 29 September 1919, folder A, box 1, Jerome C. Hunsaker Papers, Archives, National Air and Space Museum (hereafter cited as JCHP/NASM).
7. Henri Loyrette, *Gustave Eiffel,* trans. Rachel Gomme and Susan Gomme (New York: Rizzoli International Publications, 1985), 96–98, 206–11; J. C. Hunsaker, "Europe's Facilities for Aeronautical Research—I," *Flying* 3 (April 1914): 93; Anderson, *History of Aerodynamics,* 268–79.
8. Eiffel to JCH, 23 May 1912; JCH to Eiffel, 13 June 1912; JCH to Capt. Washington I. Chambers, 26 June 1912; Earle L. Ovington to JCH, 28 June 1912; all in folder 12, box 2, JCHP/MIT.
9. Capt. Washington I. Chambers to JCH, 29 June 1912; W. O. Wiley to JCH, 2 July 1912; Ferris Grenslet to JCH, 17 July 1912; JCH to Eiffel, 24 July 1912; W. M.

Meredith to Ferris Grenslet, 7 August 1912; JCH to W. M. Meredith, 27 August 1912; all in folder 12, box 2, JCHP/MIT.

10. Bird, *Jerome Clarke Hunsaker,* 166–67; Jerome C. Hunsaker Jr., interview by Earl A. Thornton, 12 November 1996, Lincoln, Mass., p. 11; Eiffel to JCH, 18 September 1912; JCH to Eiffel, 15 October 1912; JCH to Constable and Co., 23 October 1912; JCH to Eiffel, 16 November 1912; JCH to W. M. Meredith, 26 February 1913; Roger Price (Houghton Mifflin) to JCH, 12 May 1913; W. M. Meredith to JCH, 10 July 1913; all in folder 12, box 2, JCHP/MIT.

11. *Aeronautics* 6 (June 1913): 234; *Aeronautical Journal* 17 (July 1913): 203–4; advertisement in *Aeronautics* 13 (July 1913): 40.

12. Archibald D. Turnbull and Clifford L. Lord, *History of United States Naval Aviation* (New Haven: Yale University Press, 1949), 16; Alex Roland, *Model Research: The National Advisory Committee for Aeronautics, 1915–1958,* 2 vols. (Washington, D.C.: National Aeronautics and Space Administration, 1985), 1:5–8, 11–12.

13. Roland, *Model Research,* 1:17.

14. JCH to Rear Adm. R. M. Watt, 22 April 1913, folder 29, box 2, JCHP/MIT.

15. Roland, *Model Research,* 1:324.

16. JCH Service Record, file 6733, NPRC; *New York Times,* 28 July 1913; *Times* (London), 6 August 1913; JCH, "Family Narrative, 1909–1916," box 1, folder 5, JCHP/MIT; Hunsaker interview, April 1960, OHRO, Columbia University, pp. 22–23; JCH, Report on Facilities for Aeronautical Research in England, France, and Germany, no date, pp. 12–16, folder 3, box 3, JCHP/MIT. Another copy of the report is in Hunsaker, J. C., Early Papers, 1909–1932 folder, box 12, Hunsaker Biography File, Records of the National Aeronautics and Space Administration, Record Group 255, National Archives (hereafter cited as Records of NASA, RG 255, NA).

17. JCH, Report on Facilities for Aeronautical Research, pp. 16–20, folder 3, box 3, JCHP/MIT; Hunsaker, "Europe's Facilities for Aeronautical Research—I," *Flying* 3 (April 1914): 93; A. F. Zahm, *Report on European Aeronautical Laboratories,* Smithsonian Miscellaneous Collections, vol. 62, no. 3 (Washington, D.C.: Smithsonian Institution, 1914), 15–17.

18. Paul A. Hanle, *Bringing Aerodynamics to America* (Cambridge: MIT Press, 1982), 43–50; Anderson, *History of Aerodynamics,* 257–60; JCH, Report on Facilities for Aeronautical Research, pp. 27–29, folder 3, box 3, JCHP/MIT; J. C. Hunsaker, "Europe's Facilities for Aeronautical Research—II," *Flying* 3 (May 1914): 108; Zahm, *Report on European Aeronautical Laboratories,* 18–23.

19. Hunsaker, "Europe's Facilities for Aerodynamic Research—II," 108; Hunsaker interview, April 1960, OHRO, Columbia University, pp. 24–27; JCH to W. I. Chambers, 20 October 1913, folder 3, box 3, JCHP/MIT.

20. Zahm, *Report on European Aeronautical Laboratories,* 23; JCH to August von Parseval, 8 April 1920; JCH to Peter Reichenheim, 8 April 1920; JCH to Major G. J. F. von Tschudi, 8 April 1920; JCH to Dr. Fr. Ahlborn, 8 April 1920; all in folder G, box 1, JCHP/NASM.

21. Hunsaker, "Europe's Facilities for Aeronautical Research—II," 108–9; Zahm, *Report on European Aeronautical Laboratories,* 4–10; JCH, Report on Facilities for Aeronautical Research, pp. 1–6, folder 3, box 3, JCHP/MIT; Hunsaker interview, April 1960, OHRO, Columbia University, pp. 28–29; Thornton, "MIT, Jerome C. Hunsaker, and the Origins of Aeronautical Engineering," 311–12.

22. Hunsaker, "Europe's Facilities for Aeronautical Research—II," 109; Zahm, *Report on European Aeronautical Laboratories,* 10–12; JCH, Report on Facilities for Aeronautical Research, pp. 7–12, folder 3, box 3, JCHP/MIT.

23. JCH to W. I. Chambers, 20 October 1913, folder 3, box 3, JCHP/MIT; JCH, "The Present Status of Air-Ships in Europe," *Journal of the Franklin Institute* 177 (June 1914): 597–639.

24. For an excellent study of the Wrights' aerodynamic research, see Peter L. Jakab, *Visions of a Flying Machine: The Wright Brothers and the Process of Invention* (Washington, D.C.: Smithsonian Institution Press, 1990). JCH quotation from JCH, "Aeronautical Research," 20 December 1913, folder 29, box 2, JCHP/MIT.

25. Hanle, *Bringing Aerodynamics to America,* 123–35, 156; Anderson, *History of Aerodynamics,* 295.

26. J. C. Hunsaker, "Aeronautical Engineering, MIT, 1909–1916," p. 11, folder 29, box 2, JCHP/MIT. See also *Tech* 33 (30 December 1913): 1–4.

27. *Times* (London), 5 November 1913, 14 November 1913; *Tech* 33 (30 December 1913): 1,4; Hunsaker interview, April 1960, OHRO, Columbia University, p. 17; Prescott, *When MIT Was "Boston Tech,"* 284; JCH, "Aeronautical Engineering, MIT, 1909–1916," p. 12, folder 29, box 2, JCHP/MIT.

28. Thornton, "MIT, Jerome C. Hunsaker, and the Origins of Aeronautical Engineering," 311.

29. Ober, "Story of Aeronautics at MIT," 7–8; Hunsaker interview, April 1960, OHRO, Columbia University, pp. 31–32; J. C. Hunsaker, "Scientific Aeronautic Research: The New Aerodynamic Laboratory of the Massachusetts Institute of Technology," *Scientific American, Supplement,* no. 2057 (5 June 1915): 364–65.

30. Ober, "Story of Aeronautics at MIT," 8–9; Hunsaker interview, April 1960, OHRO, Columbia University, pp. 33–34.

31. Frank Cunningham, *Skymaster: The Story of Donald Douglas* (Philadelphia: Dorrance, 1943), 56.

32. Thornton, "MIT, Jerome C. Hunsaker, and the Origins of Aeronautical Engineering," 313; JCH, "Aeronautical Engineering, MIT, 1909–1916," p. 15, folder 29, box 2, JCHP/MIT.

33. Jerome C. Hunsaker, *Dynamical Stability of Aeroplanes,* Smithsonian Miscellaneous Collections, vol. 62, no. 5 (Washington, D.C.: Smithsonian Institution, 1916), 1–7, 78. For a summary and analysis of Hunsaker's work, see Thornton, "MIT, Jerome C. Hunsaker, and the Origins of Aeronautical Engineering," 313–14.

34. Roland, *Model Research,* 1:21–25; JCH, "Aeronautical Engineering, MIT, 1909–1916," p. 20, folder 29, box 2, JCHP/MIT; J. C. Hunsaker, "Experimental Analysis of Inherent Longitudinal Stability for a Typical Biplane," in *First Annual Report of*

the National Advisory Committee for Aeronautics, 1915 (Washington, D.C.: U.S. National Advisory Committee for Aeronautics, 1916), 13–14, 25–51.

35. Thornton, "MIT, Jerome C. Hunsaker, and the Origins of Aeronautical Engineering," 314.

36. Walter G. Vincenti, *What Engineers Know and How They Know It: Analytical Studies from Aeronautical History* (Baltimore: Johns Hopkins University Press, 1990), 4, xx, 62–63, 72, 233.

37. Anderson, *History of Aerodynamics,* 251–56, 293–94.

38. JCH, "Family Narrative, 1909–1916," pp. 4–5, folder 5, box 1, JCHP/MIT.

3. The War in Washington

1. Richard K. Smith, *First Across! The U.S. Navy's Transatlantic Flight of 1919* (Annapolis: Naval Institute Press, 1973), 16, 21; Jerome C. Hunsaker, "The Material Development of Naval Aircraft," vol. 10, "Aircraft Division, Bureau of Construction and Repair, Personnel and Organization for Work, July, 1916, to September, 1921," p. 2, folder 26, box 3, JCHP/MIT.

2. Smith, *First Across!* 15–18; Hunsaker interview, April 1960, OHRO, Columbia University, pp. 46–49.

3. JCH, "The Material Development of Naval Aircraft," vol. 10, "Aircraft Division, Bureau of Construction and Repair, Personnel and Organization for Work," p. 5, folder 26, box 3, JCHP/MIT.

4. Ibid., pp. 1–2; JCH, "Notes for the Chief's Hearings: School Aeroplanes—Contract No. 116," no date, folder C, box 1, JCHP/NASM; Turnbull and Lord, *History of United States Naval Aviation,* 46–47; U.S. Cong., House, *Hearings before the Committee on Naval Affairs on Estimates Submitted by the Secretary of the Navy, 1916,* 2 vols., 64th Cong, 1st sess. (Washington, D.C.: Government Printing Office, 1916), 1:3596; memo, JCH to BuC&R, 11 August 1916, folder C, box 1, JCHP/NASM.

5. JCH, "Notes for the Chief's Hearings: School Aeroplanes— Contract No. 116," no date, folder C, box 1, JCHP/NASM; JCH to Lee M. Pearson, 9 February 1962, folder 7, box 3, JCHP/MIT.

6. Clifford L. Lord, "The History of Naval Aviation, 1898–1939," Office of the Deputy Chief of Naval Operations (Air) (Washington, D.C.: Naval Aviation History Unit, 1946), 220; memo, JCH to Chief, BuC&R, 18 May 1917; memo, JCH to Chief, BuC&R, 22 May 1917; memo, W. D. Clark to Chief, BuC&R, 26 June 1917; all in folder C, box 1, JCHP/NASM.

7. Memo, JCH to Taylor, 18 May 1917, folder C, box 1, JCHP/NASM.

8. William F. Trimble, *Wings for the Navy: A History of the Naval Aircraft Factory, 1917–1956* (Annapolis: Naval Institute Press, 1990), 1–39.

9. Lord, "History of Naval Aviation," 238; William O. Shanahan, "Procurement of Naval Aircraft, 1907–1939," Deputy Chief of Naval Operations (Air) (Washington, D.C.: Naval Aviation History Unit, 1946), 17:62–65; JCH, memo, "Inspection

of Aeroplanes," attached to memo, JCH to Chief, BuC&R, 13 December 1916, folder C, box 1, JCHP/NASM.

10. Memo, JCH to Chief, BuC&R, 13 December 1916, folder C, box 1, JCHP/NASM; JCH memo, "Inspection of Aeroplanes," attached to ibid.; *The National Cyclopedia of American Biography* (New York: James T. White, 1961), 43:103; Westervelt (at Buffalo) to BuC&R, 2 May 1917, file 401-Z-0, vol. 1, box 286, Bureau of Aeronautics, General Correspondence initiated in the Bureau of Construction and Repair, 1917–1925, Record Group 72, National Archives (hereafter cited as BuAer, Gen. Corresp. initiated in BuC&R, 1917–1925, RG 72, NA); memo, JCH to Chief, BuC&R, 23 May 1917, folder C, box 1, JCHP/NASM.

11. JCH, "The Material Development of Naval Aircraft," vol. 5, "Non-Rigid Airships for the Navy," pp. 1–2, file 602-0, box 5, Office Services Division, Records of the Bureau of Aeronautics, Record Group 72, National Archives (hereafter cited as OSD, BuAer Records, RG 72, NA); Turnbull and Lord, *History of United States Naval Aviation,* 50–51; memo, JCH to Taylor, 11 August 1916, folder C, box 1, JCHP/NASM.

12. Hunsaker, "Present Status of Air-Ships in Europe," pp. 597–639; Turnbull and Lord, *History of United States Naval Aviation,* 83–84; JCH, "The Material Development of Naval Aircraft," vol. 5, "Non-Rigid Airships for the Navy," pp. 2–3, file 602-0, box 5, OSD, BuAer Records, RG 72, NA; J. C. Hunsaker, "The Navy's First Airships," *United States Naval Institute Proceedings* 45 (August 1919): 1349–50.

13. Hunsaker, "Navy's First Airships," 1350–54.

14. Taylor to SecNav, 6 January 1917, 13311-A120; SecNav to BuC&R and BuStEng, 4 February 1917, Mat. 5-D 248-17; both in file 0-ZN-12, vol. 1, box 412, BuAer, Gen. Corresp. initiated in BuC&R, 1917–1925, RG 72, NA; JCH, "The Material Development of Naval Aircraft," vol. 5, "Non-Rigid Airships," p. 3, file 602-0, box 5, OSD, BuAer Records, RG 72, NA; Turnbull and Lord, *History of United States Naval Aviation,* 84.

15. Hunsaker, "Navy's First Airships," 1356–57; Curtiss Buffalo to BuC&R, 11 February 1917; abstract of conference at BuC&R on nonrigid airships, 12 February 1917; both in file O-ZN-12, vol. 1, box 412, BuAer, Gen. Corresp. initiated in BuC&R, 1917–1925, RG 72, NA.

16. Abstract of conference at BuC&R on nonrigid airships, 12 February 1917, file O-ZN-12, vol. 1, box 412, BuAer, Gen. Corresp. initiated in BuC&R, 1917–1925, RG 72, NA; Hunsaker, "Navy's First Airships," 1356–57.

17. Specifications for the Construction of a Non-rigid Naval Dirigible for Coast Patrol, 6 February 1917, file O-ZN-12, vol. 1, box 412, BuAer, Gen. Corresp. initiated in BuC&R, 1917–1925, RG 72, NA; Hunsaker, "Navy's First Airships," 1358.

18. Gordon Swanborough and Peter M. Bowers, *United States Navy Aircraft since 1911* (Annapolis: Naval Institute Press, 1990), 571–72; Hunsaker, "Navy's First Airships," 1359–62.

19. Eric Shatzberg, *Wings of Wood, Wings of Metal: Culture and Technical Choice in*

American Airplane Materials, 1914–1945 (Princeton: Princeton University Press, 1999), 3–6, 11–15, 59–61; see also Schatzberg, "Ideology and Technical Choice: The Decline of the Wooden Airplane in the United States, 1920–1945," *Technology and Culture* 35 (January 1994): 54–55.

20. JCH, "The Material Development of Naval Aircraft," vol. 6, "Rigid Airships," p. 2, file 602-0, box 5, OSD, BuAer Records, RG 72, NA; Daniels, "Duralumin," 751–52; Schatzberg, *Wings of Wood,* 26.

21. Knox to JCH, 12 August 1916; Harold D. Sill to Taylor, 12 August 1916; both in file 160-Z-12, vol. 1, box 431, BuAer, Gen. Corresp. initiated in BuC&R, 1917–1925, RG 72, NA; Margaret B. W. Graham and Bettye H. Pruitt, *R&D for Industry: A Century of Technical Innovation at Alcoa* (Cambridge: Cambridge University Press, 1990), 160.

22. JCH, "The Material Development of Naval Aircraft," vol. 6, "Rigid Airships," pp. 3–4, file 602-0, box 5, OSD, BuAer Records, RG 72, NA; Graham and Pruitt, *R&D for Industry,* 164–65.

23. Douglas H. Robinson and Charles L. Keller, *"Up Ship!" U.S. Navy Rigid Airships, 1919–1935* (Annapolis: Naval Institute Press, 1982), 8–9.

24. Ibid.; JCH, "The Material Development of Naval Aircraft," vol. 6, "Rigid Airships," pp. 5–7, file 602-0, box 5, OSD, BuAer Records RG 72, NA.

25. Lord, "History of Naval Aviation," 329; *Aviation and Aeronautical Engineering* 2 (1 June 1917): 404; JCH to Claude A. Swanson, 11 May 1918, 601-Z-2, file 601-Z-2, box 436, BuAer, Gen. Corresp. initiated in BuC&R, 1917–1925, RG 72, NA; Irving Brinton Holley Jr., *Ideas and Weapons: Exploitation of the Aerial Weapon by the United States during World War I* (New Haven: Yale University Press, 1953), 40.

26. Turnbull and Lord, *History of United States Naval Aviation,* 110; Holley, *Ideas and Weapons,* 41–42.

27. Holley, *Ideas and Weapons,* 67; Turnbull and Lord, *History of United States Naval Aviation,* 108–10; JCH, "The Material Development of Naval Aircraft," vol. 2, "The National Advisory Committee for Aeronautics and the Aircraft Board as Affecting the Navy," p. 3, file 602-0, box 5, OSD, BuAer Records, RG 72, NA; JCH to Claude A. Swanson, 11 May 1918, 601-Z-2, file 601-Z-2, box 436, BuAer, Gen. Corresp. initiated in BuC&R, 1917–1925, RG 72, NA.

28. JCH to Claude A. Swanson, 11 May 1918, 601-Z-2, file 601-Z-2, box 436, BuAer, Gen. Corresp. initiated in BuC&R, 1917–1925, RG 72, NA; Turnbull and Lord, *History of United States Naval Aviation,* 114–15.

29. Memo, JCH to Chief, BuC&R, 12 November 1917, folder C, box 1, JCHP/NASM; JCH, "The Material Development of Naval Aircraft," vol. 4, "War Production of Aircraft by the Bureau of Construction and Repair," pp. 9–10, file 602-0, box 5, OSD, BuAer Records, RG 72, NA; Trimble, *Wings for the Navy,* 17, 20–28.

30. JCH, "The Material Development of Naval Aircraft," vol. 7, "Naval Aircraft Factory," p. 4, folder 23, box 3, JCHP/MIT; Hunsaker interview, April 1960, OHRO,

Columbia University, pp. 45–46; Jerome C. Hunsaker, "Progress in Naval Aircraft," *Society of Automotive Engineers, Transactions* 14, pt. 2 (1919): 237–38.

31. Taylor to SecNav, 8 August 1918, file 404-Z-2, box 289, BuAer, Gen. Corresp. initiated in BuC&R, 1917–1925, RG 72, NA; memo of conference, JCH to Taylor, 13 June 1918, file 601-Z-1, vol. 4, box 434A, BuAer, Gen. Corresp. initiated in BuC&R 1917–1925, RG 72, NA.

32. JCH, "The Material Development of Naval Aircraft," vol. 4, "War Production of Aircraft by the Bureau of Construction and Repair," p. 11, file 602-0, box 5, OSD, BuAer Records, RG 72, NA.

33. JCH route slip comments, 7 November 1917, on Lt. (j.g.) G. S. Gillespie to CNO, 6 August 1917, file 602-Z-1, box 436, BuAer, Gen. Corresp. initiated in BuC&R, 1917–1925, RG 72, NA; Shanahan, "Procurement of Naval Aircraft, 1909–1939," 240–41.

34. JCH to Claude A. Swanson, 11 May 1918, 601-Z-2, file 601-Z-2, BuAer, Gen. Corresp. initiated in BuC&R, 1917–1925, RG 72, NA; Shanahan, "Procurement of Naval Aircraft, 1909–1939," 112–14.

35. Hunsaker, "Navy's First Airships," 1364–65; Taylor to CNO and Chief, Bureau of Steam Engineering, 6 February 1918, O-ZN-17, 502-Z-1, file O-ZN-30, vol. 1, box 416, BuAer, Gen. Corresp. initiated in BuC&R, 1917–1925, RG 72, NA; Daniels to BuC&R and Bureau of Steam Engineering, 18 March 1918, Op-Air 0134-30, 16-AMB, file O-ZN-30, vol. 1, box 416, BuAer, Gen. Corresp. initiated in BuC&R, 1917–1925, RG 72, NA.

36. Hunsaker, "Navy's First Airships," 1365–67.

37. Taylor to SecNav, 13 May 1918, O-ZN-17; JCH to Lt. D. T. Hood, Supt. Const. Aircraft, Akron, 8 June 1918, O-ZN-17; memo, C. P. Burgess to JCH, 27 August 1918, O-ZN-17(A); all in file O-ZN-30, vol. 1, box 416, BuAer, Gen. Corresp. initiated in BuC&R, 1917–1925, RG 72, NA.

38. JCH to Garland Fulton (from Akron), 14 September 1918, file O-ZN-30; memo. of telephone conversation, Starr Truscott with Ens. Sargent, 9 October 1918, N-3509, O-Z-51(A); both in vol. 1, box 416, BuAer, Gen. Corresp. initiated in BuC&R, 1917–1925, RG 72, NA; Hunsaker, "Navy's First Airships," 1367.

39. JCH handwritten notes, 14 October 1917, on memo. for BuC&R annual report, 11 September 1918, file 601-Z-2, box 436, BuAer, Gen. Corresp. initiated in BuC&R, 1917–1925, RG 72, NA; Hunsaker interview, April 1960, OHRO, Columbia University, p. 54; JCH, "The Material Development of Naval Aircraft," vol. 10, "Aircraft Division, Bureau of Construction and Repair, Personnel and Organization for Work," pp. 4–5, enclosure U, folder 26, box 3, JCHP/MIT; Clark G. Reynolds, *Admiral John H. Towers: The Struggle for Naval Air Supremacy* (Annapolis: Naval Institute Press, 1991), 117.

40. JCH, "The Material Development of Naval Aircraft," vol. 10, "Aircraft Division, Bureau of Construction and Repair, Personnel and Organization for Work," pp. 4–7, folder 26, box 3, JCHP/MIT; JCH, "The Material Development of Naval Air-

craft," vol. 4, "War Production of Aircraft by the Bureau of Construction and Repair," p. 2, file 602-0, box 5, OSD, BuAer Records, RG 72, NA; Hunsaker interview, April 1960, OHRO, Columbia University, p. 40; for de Florez, see *New York Times,* 6 December 1962.

41. JCH to Louis M. Ream, 11 August 1917, folder R, box 1, JCHP/NASM; JCH, "The Material Development of Naval Aircraft," vol. 10, "Aircraft Division, Bureau of Construction and Repair, Personnel and Organization for Work," pp. 6–8, folder 26, box 3, JCHP/MIT.

42. JCH, "The Material Development of Naval Aircraft," vol. 10, "Aircraft Division, Bureau of Construction and Repair, Personnel and Organization for Work," p. 7, folder 26, box 3, JCHP/MIT; JCH, "The Material Development of Naval Aircraft," vol. 4, "War Production of Aircraft by the Bureau of Construction and Repair," p. 1, file 602-0, box 5, OSD, BuAer Records, RG 72, NA; Roland, *Model Research,* 1:75; Richard K. Smith, *The Airships* Akron *and* Macon: *Flying Aircraft Carriers of the United States Navy* (Annapolis: Naval Institute Press, 1965), 7.

43. J. C. Hunsaker, "Progress in Naval Aircraft," *Journal of the Society of Automotive Engineers* 5 (July 1919): 32; Smith, *First Across!* 18; G. C. Westervelt, H. C. Richardson, and A. C. Read, *The Triumph of the N.C's* (Garden City: Doubleday, Page, 1920), 4, 16.

44. Smith, *First Across!* 19–20; Hunsaker, "Progress in Naval Aircraft," *SAE Journal,* 32.

45. Smith, *First Across!* 20–21; Hunsaker, "Progress in Naval Aircraft," *SAE Journal,* 32; William J. Armstrong, "Dick Richardson: His Life in Aeronautics," *Naval Aviation News* (April 1977): 34–37.

46. Hunsaker interview, April 1960, OHRO, Columbia University, p. 59; Smith, *First Across!* 21–22.

47. Westervelt, Richardson, and Read, *Triumph of the N.C's,* 64–65; Smith, *First Across!* 22–23; Hunsaker, "Progress in Naval Aircraft," *SAE Journal,* 32.

48. Smith, *First Across!* 23; Hunsaker, "Progress in Naval Aircraft," *SAE Journal,* 38; Westervelt, Richardson, and Read, *Triumph of the N.C's,* 95–96.

49. Hunsaker, "Progress in Naval Aircraft," *SAE Journal,* 38–40; Westervelt, Richardson, and Read, *Triumph of the N.C's,* 72, 104.

50. JCH, "NC Transatlantic Flight (1919)," pp. 1, 13, folder 30, box 2, JCHP/MIT; Smith, *First Across!* 163–89.

51. David F. Noble, *America by Design: Science, Technology, and the Rise of Corporate Capitalism* (New York: Knopf, 1977), xiii.

4. Peacetime Projects and Reorganization

1. JCH, "The Material Development of Naval Aircraft," vol. 4, "War Production of Aircraft by the Bureau of Construction and Repair," pp. 3–8, file 602-0, box 5, OSD, BuAer Records, RG 72, NA.

2. Record of Proceedings of a Naval Examining Board convened at the Navy Yard, Philadelphia, in the case of Lt. Jerome C. Hunsaker, U.S. Navy, 1 March 1918, vol. 762, Hunsaker, Jerome, Proceedings of Naval and Marine Examining Boards, box 553, JAG Records (Navy), RG 125, NA.

3. Taylor to Office of Naval Intelligence, 4 November 1918, 401-Z-10 506-Z-1(A) O-ZN-12; telegram, Sims to BuC&R, 8 November 1918, S-2-87 X-27; telegram, Taylor to Sims, 12 November 1918, S-Z-87(A); all in file 500-Z-1, box 290, BuAer, Gen. Corresp. initiated in BuC&R, 1917–1925, RG 72, NA.

4. Telegram, Litchfield to JCH, 14 November 1918, file 500-Z-5, box 290, BuAer, Gen. Corresp. initiated in BuC&R, 1917–1925, RG 72, NA; JCH to Alice, 22 November 1918, folder 8, box 1, JCHP/MIT.

5. JCH to Alice, 25 November 1918, folder 8, box 1, JCHP/MIT; Capt. Ralph S. Barnaby (Ret.), interview by the author, Philadelphia, 12 October 1983.

6. JCH to Brown (?), 7, 16 December 1918, folder 8, box 1; JCH Report No. 1, 14 December 1918, folder 9, box 1; both in JCHP/MIT; Report of the Aircraft Section of the Allied Naval Armistice Commission, 20 December 1918, A-1-q(q/lc), no. 11388, box 94, Office of Naval Intelligence, Intelligence Division—Naval Attaché Reports, 1886–1939, Records of the Deputy Chief of Naval Operations, 1882–1954, Record Group 38, National Archives (hereafter cited as Attaché Reports, 1886–1939, DCNO Records, RG 38, NA); Hunsaker interview, April 1960, OHRO, Columbia University, p. 53.

7. JCH to Taylor, 21 December 1918, folder 8, box 1; memo, JCH to Roosevelt, 31 January 1919, folder 8, box 1; JCH to Air Commodore E. A. Masterman, 21 August 1919, folder 2, box 4; all in JCHP/MIT; JCH, "The Material Development of Naval Aircraft," vol. 6, "Rigid Airships," pp. 12, 16, file 602-0, box 5, OSD, BuAer Records, RG 72, NA; Hunsaker interview, April 1960, OHRO, Columbia University, p. 76.

8. JCH, Report 91, 15 February 1919, folder 9, box 1, JCHP/MIT; Capt. Ralph S. Barnaby (Ret.), interview by the author, Philadelphia, 12 October 1983.

9. JCH, "The Material Development of Naval Aircraft," vol. 9, "Formation of the Bureau of Aeronautics of the Navy Department: Notes by Comdr. J. C. Hunsaker," p. 6, Historical Papers Submitted, folder 25, box 3, JCHP/MIT; Hunsaker interview, April 1960, OHRO, Columbia University, p. 104; *New York Times,* 1 March 1919.

10. Fred C. Dickey Jr., "U.S. Navy Ship Plane Units," *Journal of the American Aviation Historical Society* 10 (Summer 1965): 121–26; Charles M. Melhorn, *Two-Block Fox: The Rise of the Aircraft Carrier, 1911–1929* (Annapolis: Naval Institute Press, 1974), 29–44.

11. Hearings before the General Board of the Navy, 1919 (micro. roll 3), 3:1242–47; Hearings before the General Board of the Navy, 1920 (micro. roll 3), 1:1, 11–12.

12. Taylor to CNO, 19 April 1921, 0-ZG-1, RNAF-Z-51(A), file 449, General Board Records, Operational Archives Branch, Naval Historical Center (now in the National Archives); Contract 54109, 30 June 1921, vol. 1, box 127, BuAer, Gen. Cor-

resp. initiated in BuC&R, 1917–1925, RG 72, NA; JCH to Comdr. E. S. Land, 18 July 1922, folder 5, file L, box 3, JCHP/NASM; Swanborough and Bowers, *United States Navy Aircraft since 1911,* 368–70.

13. Hearings before the General Board of the Navy, 1919 (micro. roll 2), 2:583–92, 603–18, 622–43.

14. Ibid., 2:704–27, 746–47.

15. Robinson and Keller, *"Up Ship!"* 11.

16. Ibid., 12; JCH Service Record, file 6733, NPRC.

17. JCH to Comdr. L. B. McBride, 29 August 1919; JCH to McBride, 23 October 1919; both in folder 2, box 4, JCHP/NASM; Robinson and Keller, *"Up Ship!"* 17–19, 30–31.

18. Robinson and Keller, *"Up Ship!"* 5, 18–19, 29; JCH, "The Material Development of Naval Aircraft," vol. 6, "Rigid Airships," p. 13, file 602-0, box 5, OSD, BuAer Records, RG 72, NA.

19. Robinson and Keller, *"Up Ship!"* 50–51, 53–56; JCH, "The Material Development of Naval Aircraft," vol. 6, "Rigid Airships," p. 14, file 602-0, box 5, OSD, BuAer Records, RG 72, NA; JCH to Gen. E. A. Masterman, 21 August 1919, folder 2, box 4, JCHP/NASM.

20. Hearings before the General Board of the Navy, 1919 (micro. roll 2), 1:220–33.

21. Ibid., 1:485–86, 492–99.

22. JCH, "The Material Development of Naval Aircraft," vol. 9, "Formation of the Bureau of Aeronautics of the Navy Department: Notes by Commander J. C. Hunsaker," pp. 1–4, Historical Papers Submitted, folder 25, box 3, JCHP/MIT.

23. Hearings before the General Board of the Navy, 1919 (micro. roll 2), 1:897–99, 908–9.

24. Ibid., 1:899, 902.

25. Memo, JCH to Taylor, 18 July 1919, file 602-2, box 308, BuAer, Gen. Corresp. initiated in BuC&R, 1917–1925, RG 72, NA.

26. Shanahan, "Procurement of Naval Aircraft, 1907–1939," 257; Turnbull and Lord, *History of United States Naval Aviation,* 170, 186–87; U.S. Cong., House, *United Air Service. Hearing before a Subcommittee of the Committee on Military Affairs,* 66th Cong., 2d sess. (Washington, D.C.: Government Printing Office, 1919), 42–117.

27. Ibid., 451–53.

28. JCH to Comdr. Ralph T. Hanson, 22 December 1919, folder 2, box 3, JCHP/NASM.

29. Roland, *Model Research,* 1:54; JCH, "The Material Development of Naval Aircraft," vol. 9, "Formation of the Bureau of Aeronautics of the Navy Department: Notes by Commander J. C. Hunsaker," pp. 6–8, Historical Papers Submitted, folder 25, box 3, JCHP/MIT; Craven to CNO, 8 September 1920, Op-15-GB, file 3084-D-1, box 161, BuAer, Gen. Corresp. initiated in BuC&R, 1917–1925, RG 72, NA.

30. Adm. D. W. Taylor to P. A. S. Franklin, 7 May 1920, folder 2, box 2; JCH to E. Blough, 15 May 1920, folder 1, box 1; Land to Taylor, 12 June 1920, folder 2,

box 3; JCH to Taylor, 17 June 1920, folder 2, box 3; all in JCHP/NASM; *New York Times,* 19 May 1920.

31. Jerome C. Hunsaker, "Naval Architecture in Aeronautics," *Aeronautical Journal* 24 (July 1920): 321–33.

32. Samuel W. Stratton to JCH, 17 May 1920, folder 2, box 3; JCH to Joseph S. Ames, 1 September 1920, folder 2, box 3; JCH to Director of Naval Intelligence (undated, probably August 1920), folder 1, box 23; JCH to Taylor, 6 August 1920, folder 2, box 3; JCH to Joseph S. Ames, 1 September 1920, folder 2, box 3; all in JCHP/NASM.

33. James R. Hansen, *Engineer in Charge: A History of the Langley Aeronautical Laboratory, 1917–1958* (Washington, D.C.: National Aeronautics and Space Administration, 1987), 73–75. See also James R. Hansen, "The Revolt against Max Munk" (undated ms.), 3–5 (courtesy of James R. Hansen).

34. Hansen, *Engineer in Charge,* 69–72; Munk to JCH, 7 October 1920, folder 2, box 4, JCHP/NASM.

35. JCH to Munk, 2 November 1920; Munk to JCH, 2 January 1921; both in folder 2, box 4, JCHP/NASM; Hansen, *Engineer in Charge,* 75.

36. JCH to Director of Naval Intelligence (undated, probably August 1920), folder 1, box 23; JCH to Taylor, 6 August 1920, folder 2, box 3; JCH to von Parseval, 8 April 1920, folder 4, box 2; all in JCHP/NASM.

37. Richard K. Smith, "A Douglas Decision," *Aerospace Historian* 15 (Autumn 1968): 5–6; Hunsaker interview, April 1960, OHRO, Columbia University, pp. 71–72.

38. C. R. Roseberry, *Glenn Curtiss: Pioneer of Flight* (Garden City, N.Y.: Doubleday, 1972), 329–30, 409–10; Naval Orders, July 1916–November 1918, folder 15, box 3, JCHP/MIT; Hunsaker interview, April 1960, OHRO, Columbia University, pp. 90–91; Curtiss to JCH, 26 November 1920; JCH to Curtiss, 29 November 1920; both in folder 3, file C, box 1, JCHP/NASM.

39. JCH, "The Material Development of Naval Aircraft," vol. 9, "Formation of the Bureau of Aeronautics of the Navy Department: Notes by Commander J. C. Hunsaker," p. 8, Historical Papers Submitted, folder 25, box 3, JCHP/MIT; Melhorn, *Two-Block Fox,* 60–61; U.S. Cong., House, *Hearings before the Committee on Naval Affairs . . . on Sundry Legislation Affecting the Naval Establishment, 1920–1921,* 66th Cong., 3d sess. (Washington, D.C.: Government Printing Office, 1921), 116–18, 217, 221, 651, 703; William F. Trimble, *Admiral William A. Moffett: Architect of Naval Aviation* (Washington, D.C.: Smithsonian Institution Press, 1994), 80–81; Hunsaker interview, April 1960, OHRO, Columbia University, p. 67.

40. JCH to Comdr. Lewis M. Maxfield, 24 January 1921, folder 2, box 4, JCHP/NASM; JCH, "The Material Development of Naval Aircraft," vol. 9, "Formation of the Bureau of Aeronautics of the Navy Department: Notes by Commander J. C. Hunsaker," p. 8, Historical Papers Submitted, folder 25, box 3, JCHP/MIT; Trimble, *Moffett,* 76–80.

41. Burke Davis, *The Billy Mitchell Affair* (New York: Random House, 1967), 77–112; BuNav to JCH, 15 July 1921, N-313-PG, folder 1, box 6, JCHP/NASM.

42. Hunsaker interview, April 1960, OHRO, Columbia University, p. 52; Turnbull and Lord, *History of United States Naval Aviation,* 200–1; JCH to Maj. B. D. Foulois, 14 December 1921, folder 2, box 2, JCHP/NASM.

43. Robinson and Keller, *"Up Ship!"* 31, 37, 43–46; JCH to W. C. Geer, 8 September 1921, folder 4, box 2, JCHP/NASM.

44. Robinson and Keller, *"Up Ship!"* 47–48; C. P. Burgess to JCH, 27 November 1921, folder 2, box 1, JCHP/NASM.

45. Circular Letter 1, 10 August 1921, in World War II Administrative History, BuAer, vol. 1, Background (Washington, D.C.: Department of the Navy, 1957), pp. 246–51; memo, JCH to Moffett, 3 August 1921, file 602-4, box 308, BuAer, Gen. Corresp. initiated in BuC&R, 1917–25, RG 72, NA; JCH to Comdr. E. S. Land, 2 July 1921, folder 5, box 3, JCHP/NASM.

46. JCH to Comdr. E. S. Land, 2 July 1921, folder 5, box 3, JCHP/NASM; Circular Letter 1, 10 August 1921, in World War II Administrative History, BuAer, vol. 1, Background, pp. 246–51; JCH, "The Material Development of Naval Aircraft," vol. 9, "Formation of the Bureau of Aeronautics of the Navy Department: Notes by Commander J. C. Hunsaker," p. 8, Historical Papers Submitted, folder 25, box 3, JCHP/MIT; William A. Moffett, "Organization and Function of Naval Aviation," *Aviation* 13 (28 August 1922): 248–49.

47. JCH to Land, 21 December 1920; Land to JCH, 3 March 1921; both in folder 21, box 1, JCHP/MIT; Hunsaker interview, April 1960, OHRO, Columbia University, p. 68.

48. JCH to Lt. T. P. Wright, 6 October 1921, folder 5, box 6; JCH to Lt. Comdr. W. W. Webster, 1 February 1922, folder 6, box 6; both in JCHP/NASM.

49. BuNav to JCH, 15 December 1921, N3130AB 5-615; BuNav to JCH, 19 December 1921, 6615-184 N322-FJS-GW; both in 00/Hunsaker, Jerome C., box 4120, file 00, OSD, BuAer, Gen. Corresp., 1925–42, RG 72, NA; Bird, *Hunsaker Family History,* 72; JCH to Lt. Comdr. W. D. Brereton, 8 October 1921, folder 2, box 1; JCH to Ralph Hanson, 26 April 1923, folder 2, box 3; both in JCHP/NASM.

5. Service at Home and Abroad

1. JCH to C. I. R. Campbell, 15 April 1920, folder 3, box 1; JCH to Lewis M. Maxfield, 24 January 1921, folder 2, box 4; JCH to Horace Dyer, 25 June 1921, folder 5, box 1; all in JCHP/NASM; Robinson and Keller, *"Up Ship!"* 54–55; Trimble, *Wings for the Navy,* 57.

2. JCH to C. P. Burgess, 18 March 1922, folder 2, box 1, JCHP/NASM.

3. Ibid.

4. Ibid.; JCH to Garland Fulton, 16 January 1923, folder 3, box 2, JCHP/NASM; Robinson and Keller, *"Up Ship!"* 58–59.

5. JCH, "The Material Development of Naval Aircraft," vol. 6, "Rigid Airships," p. 15, file 602-0, box 5, OSD, BuAer Records, RG 72, NA; Edwin Denby (SecNav) to President, 11 October 1922, Aer-A-FD, 800-31, 600-3, 00/Hunsaker,

Jerome C., box 4120, file 00, OSD, BuAer, Gen. Corresp., 1925–1942, RG 72, NA.

6. Robinson and Keller, *"Up Ship!"* 64–67; JCH to Chief, BuAer, 12 December 1922; A. W. Johnson to Chief, BuNav, 24 February 1923, Aer-F-3-BGS, 800-3; BuAer to INA, Friedrichshafen, 4 April 1923, Aer-M-1-QL, Aero351BU-22, 800-31, 801-18; Moffett to Chief, BuNav, 23 May 1923, Aer-M-1-CK, 800-3; all in 00/Hunsaker, Jerome C., box 4120, file 00, OSD, BuAer, Gen. Corresp., 1925–1942, RG 72, NA.

7. Robinson and Keller, *"Up Ship!"* 118–24; JCH to Lt. Comdr. W. W. Webster, 1 February 1922; Webster to JCH, 4 February 1922; both in folder 6, box 6, JCHP/NASM.

8. JCH to Fulton, 16 January 1923, folder 3, box 2, JCHP/NASM.

9. *Times* (London), 18 June, 27 July 1923; *New York Times,* 3 August 1923; Bureau of Aeronautics, *Weekly News Letter* 3 (13 June 1923): 5; Hunsaker interview, April 1960, OHRO, Columbia University, p. 77; Robinson and Keller, *"Up Ship!"* 122; Chief, BuNav to JCH, 7 June 1923, N-312-E, 00/Hunsaker, Jerome C., box 4120, file 00, OSD, BuAer, Gen. Corresp., 1925–1942, RG 72, NA; Towers to JCH, 20 September 1923, folder 1, box 6; JCH to Griffith Brewer, 16 August 1923, folder 2, box 1; both in JCHP/NASM.

10. Moffett to Chief, BuNav, 14 August 1923, Aer-F-3-BGS, 800-31; Denby to JCH, 20 September 1923, Aer-GB, 800-31, OR-4; both in 00/Hunsaker, Jerome C., box 4120, file 00, OSD, BuAer, Gen. Corresp., 1925–1942, RG 72, NA; Upson to JCH, 7 September 1923, folder 2, box 6, JCHP/NASM.

11. Naval Orders, July 1916–November 1918, folder 15, box 3, JCHP/MIT; JCH to Westervelt, 19 March 1920, file W, box 6, JCHP/NASM; U.S. Cong., House, *Hearings before the Committee on Naval Affairs . . . on Estimates Submitted by the Secretary of the Navy, 1920,* 2 vols., 66th Cong., 2d sess. (Washington, D.C.: Government Printing Office, 1920), 1:634–37.

12. Trimble, *Moffett,* 111–13.

13. Memo, JCH to Moffett, 21 October 1921, vol. 6, file NP 11, box 4025, BuAer, Gen. Corresp., 1925–1942, RG 72, NA.

14. Trimble, *Wings for the Navy,* 67.

15. Trimble, *Moffett,* 137–38.

16. Chief, BuAer to Chief, BuNav, 1 May 1922, Aer-F-BGS, 803-0, 800-353, 800-431; BuNav to JCH, 2 May 1922, N-312-UB-KR; both in file 00/Hunsaker, Jerome C., box 4120, file 00, OSD, BuAer, Gen. Corresp., 1925–1942, RG 72, NA; Cary T. Grayson, Physical Examination of J. C. Hunsaker for Aviation Duty, 8 May 1922, folder 8, box 2, JCHP/MIT; Moffett to Chief, BuNav, 18 May 1922, Aer-F-GB, 803-0, vol. 3, file 803-0, box 337, BuAer, Gen. Corresp. initiated in BuC&R, 1917–1925, RG 72, NA; JCH to Garland Fulton, 29 September 1922, folder 3, box 2, JCHP/NASM; Trimble, *Moffett,* 137.

17. JCH, "Technical Developments, 1920," pp. 19–20 (draft for *Aircraft Yearbook,* 1921), folder 21, box 1, JCHP/MIT; Thomas G. Foxworth, *The Speed Seekers* (New York: Doubleday, 1976), 28, 161–73, 177–83.

18. JCH, "Technical Developments, 1920," pp. 20–22 (draft for *Aircraft Yearbook,* 1921), folder 21, box 1, JCHP/MIT; Adm. D. W. Taylor to Contest Committee, Aero Club of America, 20 November 1920, folder 1, box 1, JCHP/NASM; Foxworth, *Speed Seekers,* 30–32, 173–74, 446–50.

19. Foxworth, *Speed Seekers,* 33–34, 194–98; JCH to Thurman Bane, 18 February 1922; Bane to JCH, 27 February 1922; both in folder 2, box 1, JCHP/NASM.

20. Foxworth, *Speed Seekers,* 297–98, 540, 461–62; JCH to Lt. Comdr. Newton H. White, 8 February 1923, folder 4, box 6, JCHP/NASM.

21. Foxworth, *Speed Seekers,* 34–35, 199, 210, 244–47, 494–95, 519; J. C. Hunsaker, "Notes on Aeroplane Shock Absorbers of Rubber," *Aeronautics* (London) 11 (11 October 1916): 235–39.

22. Bureau of Navigation to JCH, 29 September 1922, N-310-PG, 00/Hunsaker, Jerome C., box 4120, file 00, OSD, BuAer, Gen. Corresp., 1925–1942, RG 72, NA; JCH to Lt. Comdr. Newton H. White, 8 February 1923, folder 4, box 6, JCHP/NASM.

23. JCH to Lt. Comdr. Newton H. White, 8 February 1923, folder 4, box 6, JCHP/ NASM; Foxworth, *Speed Seekers,* 52–53, 201–4, 248–52, 472–73.

24. Foxworth, *Speed Seekers,* 35–37, 218–21, 252–55; Bureau of Navigation to JCH, 24 September 1923, N-312-E, 00/Hunsaker, Jerome C., box 4120, file 00, OSD, BuAer, Gen. Corresp., 1925–1942, RG 72, NA.

25. Roland, *Model Research,* 2:428–29, 432; Report on the Fitness of Officers, Jerome C. Hunsaker, 21 June 1922, vol. 762, Hunsaker, Jerome, Proceedings of Naval and Marine Examining Boards, box 553, JAG Records (Navy), RG 125, NA.

26. J. D. Beuret (BuC&R) to BuNav, via BuAer and ONI, 14 September 1923, 6345-A()), 10/1; Moffett to Chief, BuNav, via CNO and ONI, 20 September 1923, Aer-GB, 800-31; both in 00/Hunsaker, Jerome C., box 4120, file 00, OSD, BuAer, Gen. Corresp., 1925–1942, RG 72, NA; Bird, *Hunsaker Family History,* 72–74.

27. JCH, "The Material Development of Naval Aircraft," 4 December 1923, file 602-0, box 5, OSD, BuAer Records, RG 72, NA (another copy is in folders 25 and 26, box 3, JCHP/MIT). See also JCH to Moffett, 19 November 1922, Aer-M-1-QL, 602-0, 800-31, file 602-0, box 5, OSD, BuAer Records, RG 72, NA.

28. JCH to Director of Naval Intelligence, 17 September 1923, Aer-M-1-Wn, 602-0, 506-1, 506-2, file 602-0, box 5, OSD, BuAer Records, RG 72, NA.

29. Maj. P. L. Teed to JCH, 12 October 1923, folder 1, box 6; JCH to Maj. P. L. Teed, 5 November 1923, folder 1, box 6; JCH to George C. Westervelt, 5 November 1923, folder 5, box 6; JCH to Arthur M. Marsh, 14 November 1923, folder 1, box 4; all in JCHP/NASM; BuNav to JCH, 3 November 1923 (2d endorsement, 2 December 1923), folder 21, box 1, JCHP/MIT.

30. Wyman H. Packard, *A Century of U.S. Naval Intelligence* (Washington, D.C.: Office of Naval Intelligence and Naval Historical Center, 1996), 65; Report on the Fitness of Officers, Jerome C. Hunsaker, October 29, 1924, vol. 762, Hunsaker, Jerome, box 533, Proceedings of Naval and Marine Examining Boards, JAG Records (Navy), RG 125, NA. For information on some of Hunsaker's invest-

ments in 1924, see Arthur Marsh to JCH, 10 January 1928, folder 24, box 7, JCHP/NASM.

31. Bird, *Hunsaker Family History,* 74–75, 80; JCH to Walter Hunsaker, 21 December 1923, folder 13, box 1, JCHP/MIT; JCH to Maj. P. L. Teed, 5 November 1923, folder 1, box 6, JCHP/NASM.

32. Bird, *Hunsaker Family History,* 78–79.

33. Ibid., 75–77.

34. General information on Hunsaker's duties comes from ONI indexes, A-1-A to A-1-N, boxes 1, 2, Records of the Office of Naval Intelligence, Record Group 38, National Archives; see also Capt. Luke McNamee to ONI, 18 November 1924, register 12734-F, A-1-u, box 138, Attaché Reports, 1886-1939, DCNO Records, RG 38, NA, which includes a detailed statistical report on RAF personnel.

35. See various orders from naval attaché, London, to JCH, 7 January–28 June 1924, 23 May, 20 July 1925, folder 22, box 1, JCHP/MIT; W. W. Galbraith (by direction, Director of Naval Intelligence) to Director of Military Intelligence, 7 July 1924, Op-16-A, file 2081-715, box 843, Military Intelligence Division Corresp., 1917–1941, Records of the War Department General and Special Staffs, Record Group 165, National Archives.

36. Roland, *Model Research,* 1:75; JCH to George Lewis, 19 February 1924, J. C. Hunsaker file, Misc. Sources, NACA, NASA History Office; Lewis to JCH, 4 May 1925, Hunsaker, J. C., 1916–35 (1) (M-N) folder, box 12, Hunsaker Biog. File, Records of NASA, RG 255, NA.

37. Robinson and Keller, *Up Ship!* 130–38; Capt. C. L. Hussey to JCH, 1 September 1924, folder 22, box 1, JCHP/MIT.

38. Trimble, *Moffett,* 144–45.

39. "Great Britain: Arguments in Favor of a United Air Service as Opposed by Arguments in Favor of the Control of Naval Aviation by the Navy, 1925," Register 17399, A-1-u; McNamee to ONI, 17 October 1924, 1450, Register 17399-A; both in box 142, Attaché Reports, 1886–1939, DCNO Records, RG 38, NA.

40. Trimble, *Moffett,* 159–64.

41. C. P. Burgess to JCH, 17 September 1925, folder 12, USS *Shenandoah* Investigations, 1925, box 3, JCHP/MIT.

42. *New York Times,* 5 October 1925; Reynolds, *Towers,* 186; memo, JCH to Morrow, undated, folder 22, box 1, JCHP/MIT.

43. U.S. President's Aircraft Board, *Hearings before the President's Aircraft Board,* 4 vols. (Washington, D.C.: Government Printing Office, 1925), 4:1642–45, 1649.

44. Ibid., 4:1645–48.

45. Trimble, *Moffett,* 165–66; T. Douglas Robinson, claim of Comdr. J. C. Hunsaker, 1 September 1926, 00-Hunsaker, J. C., Gen. Corresp. of the Secretary of the Navy, RG 80, NA (courtesy of Richard W. Peuser); JCH to Morrow, 7 December 1925, folder 22, box 1, JCHP/MIT.

46. Report on the Fitness of Officers, Jerome C. Hunsaker, 30 April 1924; Report on the Fitness of Officers, Jerome C. Hunsaker, 19 April 1926; Report on the Fitness

of Officers, Jerome C. Hunsaker, 11 September 1926; all in vol. 762, Hunsaker, Jerome, Proceedings of Naval and Marine Examining Boards, box 533, JAG Records (Navy), RG 125, NA.
47. Hunsaker's decision to leave the navy appears for the first time in JCH to Dwight W. Morrow, 7 December 1925, folder 22, box 1, JCHP/MIT; Bird, *Hunsaker Family History,* 60; JCH, "Miss Porter's School, Farmington, Connecticut," folder 5, box 1, JCHP/MIT; Hunsaker interview, April 1960, OHRO, Columbia University, p. 87; JCH to SecNav, 30 June 1926, folder 22, box 1, JCHP/MIT.
48. Report on the Fitness of Officers, Jerome C. Hunsaker, 11 September 1926; Robinson to JCH, 2 September 1926; both in vol. 762, Hunsaker, Jerome, Proceedings of Naval and Marine Examining Boards, box 533, JAG Records (Navy), RG 125, NA.

6. Bell Labs

1. Richard P. Hallion, *Legacy of Flight: The Guggenheim Contribution to American Aviation* (Seattle: University of Washington Press, 1977), 26, 37; McNamee to JCH, 18 February 1926; Chief, Bureau of Navigation, to JCH, 27 April 1926; Capt. W. C. Watts to JCH, 9 July 1926; all in folder 22, box 1, JCHP/MIT. A version of this chapter appeared as "Jerome C. Hunsaker, Bell Labs, and the West Coast Model Airline," *Journal of the West* 36 (July 1997): 44–52.
2. JCH to Hugo Junkers, 8 May 1928, folder 10, box 7, JCHP/NASM; Hunsaker interview, April 1960, OHRO, Columbia University, p. 87.
3. Bureau of Navigation to JCH, 27 April 1926, first endorsement, 1 September 1926, folder 22, box 1, JCHP/MIT; misc. correspondence, folder 25, box 8, JCHP/NASM; Bird, *Hunsaker Family History,* 81; Leonard S. Reich, *The Making of American Industrial Research: Science and Business at GE and Bell, 1876–1926* (Cambridge: Cambridge University Press, 1985), 158–60, 184, 194–98.
4. Hallion, *Legacy of Flight,* 86–87; Robert J. Serling, *The Only Way to Fly: The Story of Western Airlines, America's Senior Air Carrier* (Garden City, N.Y.: Doubleday, 1976), 35–60.
5. JCH to C. C. Moseley, 23 May 1927; C. C. Moseley to JCH, June 27, 1927; both in folder 13, box 7, JCHP/NASM.
6. "A Survey of Air Transport and Its Communication Problems," 1 July 1927, in Case Survey Report, Bell Labs, prepared by P. H. Evans, 29 December 1927, folder 1, box 7; Memorandum on Aircraft Communications, folder 2, box 13; both in JCHP/NASM.
7. JCH, "Communications for Pan-American Airways," 29 August 1927, folder 1, box 7, JCHP/NASM.
8. JCH memo for file, 19 September 1927, folder 1, box 7, JCHP/NASM; Horace R. Byers, "Carl-Gustaf Arvid Rossby," in *Biographical Memoirs* (New York: National Academy of Sciences, 1960), 34:252–54.
9. JCH, memo for file, 19 September 1927, folder 1, box 7; JCH, memo for file, 27 October 1927, folder 22, box 7; both in JCHP/NASM.

10. JCH, memo for file, 19 September 1927, folder 1, box 7; JCH, memo for file, 27 October 1927, folder 22, box 7; both in JCHP/NASM.

11. JCH, memo for file, Communications for Airways (Meteorological Service), 17 November 1927, folder 1, box 7; JCH to P. H. Coolidge, 11 June 1928, folder 4, box 7; both in JCHP/NASM.

12. "Notes on Proposed Communication System for the Daniel Guggenheim Committee on Aeronautical Meteorology," 11 November 1927, folder 22, box 7; JCH to Wesley L. Smith, 11 January 1928, folder 17, box 7; both in JCHP/NASM.

13. JCH to Wesley L. Smith, 11 January 1928, folder 17, box 7, JCHP/NASM.

14. Ibid.

15. Nick A. Komons, *Bonfires to Beacons: Federal Civil Aviation Policy under the Air Commerce Act, 1926–1938* (Washington, D.C.: Department of Transportation, Federal Aviation Administration, 1978), 147–49, 155–58.

16. JCH, memo for file, 27 September 1927, folder 1, box 7; JCH to Dr. J. H. Dellinger, 13 December 1927, folder 5, box 7; JCH, memo for file, 29 March 1928, folder 1, box 7; JCH to J. C. Latham, 12 April 1928, folder 7, box 7; all in JCHP/NASM. See also D. K. Martin, "Laying a Foundation for Aircraft Communication," *Bell Laboratories Record* 7 (April 1929): 315–18.

17. JCH to Wesley L. Smith, 11 January 1928, folder 17, box 7; JCH, memo for file, 5 March 1928, folder 1, box 7; both in JCHP/NASM.

18. JCH to Edward P. Warner, 27 December 1927, folder 20, box 7, JCHP/NASM.

19. JCH to Wesley L. Smith, 11 January 1928, folder 17, box 7, JCHP/NASM; Donald R. Whitnah, *A History of the United States Weather Bureau* (Urbana: University of Illinois Press, 1961), 181–82; Serling, *Only Way to Fly,* 61–64; *Aviation* 23 (18 July 1927): 151; *Aviation* 23 (10 October 1927): 882.

20. JCH to C. G. Rossby, 10 January 1928, folder 16, box 7, JCHP/NASM.

21. JCH, memo for file, 5 March 1928; JCH, memo for file, 29 March 1928; both in folder 1, box 7, JCHP/NASM.

22. JCH to Howard (sic) Pitcairn, 21 March 1928, folder 15, box 7; JCH to C. P. Cooper, 3 May 1928, folder 4, box 7; both in JCHP/NASM; Komons, *Bonfires to Beacons,* 150–51; *New York Times,* 1 November 1928.

23. JCH to Nelson Doubleday, 23 November 1927, folder 16, box 8; JCH to Capt. M. G. Christie, 27 December 1927, folder 4, box 7; JCH to Abraham and Straus, Brooklyn, 10 April 1928, folder 23, box 7; William Bottomley to JCH, 15 May 1928, folder 24, box 7; JCH to Capt. L. B. McBride, 28 April 1928, folder 2, box 8; all in JCHP/NASM; sales brochure, 1932, from Alice Hunsaker Bird photo album, courtesy of Alice Hunsaker Bird, Henley-on-Thames, England.

24. Carl Gustaf Rossby, "Airways and the Weather," *Western Flying* 4 (April 1928): 23–24, 96.

25. Hallion, *Legacy of Flight,* 94–95.

26. JCH to J. C. Latham, 9 March 1928, folder 12, box 7; JCH to C. P. Cooper, 3 May 1928, folder 4, box 7; JCH, memo for file, 3 May 1928, folder 4, box 7; JCH to P. H. Coolidge, 25 May 1928, folder 4, box 7; all in JCHP/NASM.

27. JCH to P. H. Coolidge, 25 May 1928, folder 4, box 7, JCHP/NASM.

28. JCH to C. P. Cooper, 3 May 1928, folder 4, box 7; JCH to P. H. Coolidge, 25 May 1928, folder 4, box 7; both in JCHP/NASM.
29. C. G. Rossby, "Outline for Meteorological Service for the Airways between Los Angeles and San Francisco," folder 22, box 7, JCHP/NASM.
30. Serling, *Only Way to Fly,* 75–77; Hallion, *Legacy of Flight,* 96.
31. Hallion, *Legacy of Flight,* 93, 96–97; JCH to P. H. Coolidge, 11 June 1928, folder 4, box 7; JCH to P. H. Coolidge, 16 July 1928, folder 21, box 7; both in JCHP/ NASM.
32. Hallion, *Legacy of Flight,* 97–99.

7. Goodyear-Zeppelin

1. JCH to Paul W. Litchfield, 7 March 1928, folder 1, box 8; JCH to Litchfield, 20 June 1928, folder 12, box 7; both in JCHP/NASM.
2. JCH to Paul W. Litchfield, 20 June 1928, folder 12, box 7; JCH to Karl Arnstein, 12 July 1928, folder 2, box 7; Litchfield to JCH, 16 August 1928, folder 1, box 8; all in JCHP/NASM; Bird, *Hunsaker Family History,* 74, 80–81.
3. Smith, *Airships* Akron *and* Macon, 7.
4. Richard K. Smith, "Zeppelin Kaput!" (unpublished ms., circa 1988), 1–2 (courtesy of Richard K. Smith).
5. Ibid., 1–4.
6. Ibid., 10; Don Rose, "Bigger and Better Balloons," *Aero Digest* 14 (June 1929): 57–58, 252–56; U.S. Cong., House, *Merchants' Aircraft. Hearing before the Committee on Interstate and Foreign Commerce,* 71st Cong., 3d sess. (Washington, D.C.: Government Printing Office, 1931), 49.
7. Smith, *Airships* Akron *and* Macon, 7–18.
8. Ibid., 31; Trimble, *Moffett,* 231; W. A. Moffett, radio speech at New York Air Show, 9 May 1930; W. A. Moffett, speech at dedication of U.S. Naval Reserve Aviation Base, Miami, 8 January 1931; both in roll 13, William A. Moffett Papers, Ms. Coll. 198 (microfilm), Nimitz Library, U.S. Naval Academy.
9. JCH to John C. Latham, 22 August 1928, folder 12, box 7; JCH to Litchfield, 18 August 1928, folder 1, box 8; both in JCHP/NASM; JCH, "On the Need for an Ocean Weather Service," 16 November 1928, folder 72, box 3, JCHP/MIT.
10. Harold G. Dick and Douglas H. Robinson, *The Golden Age of the Great Passenger Airships: Graf Zeppelin and Hindenburg* (Washington, D.C.: Smithsonian Institution Press, 1985), 32–39; *New York Times,* 16 October 1928.
11. *New York Times,* 18–24, 29 October 1928.
12. JCH, Commercial Airship Proposal, March 1929, folder 2, box 5, Lighter-than-Air Society Collection, University of Akron Archives, Akron, Ohio (hereafter cited as LTA Society Coll., UA).
13. Smith, "Zeppelin Kaput!" 4; *New York Times,* 9, 10 May 1929.
14. Dick and Robinson, *Golden Age,* 39; *New York Times,* 1 September 1929.
15. *New York Times,* 1–7 September 1929.

16. Certificate of Incorporation, International Zeppelin Transport Company, 18 October 1929, folder 3, box 30, LTA Society Coll., UA; *New York Times,* 21 October 1929; U.S. Cong., House, *Merchants' Aircraft . . . Hearing,* 34.

17. Memo of Agreement, IZT, Goodyear-Zeppelin, Luftschiffbau, and National City Co., 24 March 1930, folder 3, box 30; JCH, Statement Regarding International Zeppelin Transport Corporation, 4 December 1931, pp. 1–4, folder 4, box 30; both in LTA Society Coll., UA; *Aircraft Year Book for 1931* (New York: Aeronautical Chamber of Commerce of America, Inc., 1931), 61–63; *New York Times,* 25, 30 March 1930, 29 January 1931.

18. Certificate of Incorporation of Pacific Zeppelin Transport Company, Ltd., 22 October 1929; J. C. Hunsaker, Statement Regarding Pacific Zeppelin Transport Company, Ltd., February 1932, pp. 1–3; both in folder 5, box 23, LTA Society Coll., UA; *New York Times,* 23, 29 October 1929; U.S. Cong., House, *Merchants' Aircraft . . . Hearing,* 70.

19. Hallion, *Legacy of Flight,* 67; JCH, Memorandum of Conference, 14 June 1929, Airship Institute, 1926–1929; Zook to Harry Guggenheim, 2 July 1929; both in box 4, Guggenheim Records, Library of Congress Manuscript Division, Washington, D.C. (hereafter cited as Guggenheim Records, LCMD).

20. JCH, Memorandum of Conference, 14 June 1929, Airship Institute, 1926–1929, box 4, Guggenheim Records, LCMD.

21. Ibid.; Robert A. Millikan to Theodore von Kármán, 6 July 1929, Airship Institute, 1926–1929, box 4, Guggenheim Records, LCMD; Hanle, *Bringing Aerodynamics to America,* 53–54, 63–64, 96–103, 127–30; press release, 29 October 1929, box 3, Guggenheim Records, LCMD.

22. *New York Times,* 26, 27 June 1932.

23. *New York Times,* 1 April 1930; Douglas H. Robinson, *Giants in the Sky: A History of the Rigid Airship* (Seattle: University of Washington Press, 1973), 275–77, 307–9; Hunsaker interview, April 1960, OHRO, Columbia University, p. 78; R. K. Smith, "Manhattan's Midtown Mooring Mast" (no date), folder 8, box 20, LTA Society Coll., UA.

24. Dick and Robinson, *Golden Age,* 22–23, 83–84; *New York Times,* 6 October 1930; J. C. Hunsaker, Notes on Visit to Friedrichshafen, 12 November 1930, folder 6, box 30, LTA Society Coll., UA (published as "A Visit to Friedrichshafen," *Gasbag Journal,* no. 37 (September 1998): 23–26 (courtesy of Henry Cord Meyer).

25. JCH, Notes on Visit to Friedrichshafen, 12 November 1930, folder 6, box 30, LTA Society Coll., UA; *New York Times,* 2, 5, 7 November 1930.

26. Hunsaker interview, April 1960, OHRO, Columbia University, p. 78; U.S. Cong., House, *Merchant's Aircraft . . . Hearing,* 55; J. C. Hunsaker, "Safety in Airship Transportation" (paper prepared for First International Aerial Safety Congress, Paris, 10–23 December 1930), folder 45, box 23, LTA Society Coll., UA.

27. J. C. Hunsaker, Merchant Airship Legislation, 1931–32, July 1932, pp. 1, 15, folder 4, box 5, LTA Society Coll., UA.

28. U.S. Cong., Senate, *Congressional Record,* 71st Cong., 2d sess. (Washington,

D.C.: Government Printing Office, 1930), 72(7):7104; *Aero Digest* 28 (April 26, 1930): 855; Smith, "Zeppelin Kaput!" 12; *New York Times,* 5 December 1930.

29. J. C. Hunsaker, Statement Regarding International Zeppelin Transport Corporation, 4 December 1931, pp. 20–21, figures, pp. 3–18, folder 4, box 30, LTA Society Coll., UA; U.S. Cong., House, *Merchants' Aircraft . . . Hearing,* 46–47.

30. JCH, Statement Regarding Pacific Zeppelin Transport Company, Ltd., February 1932, pp. 23–29, folder 5, box 23, LTA Society Coll., UA; U.S. Cong., Senate, *Merchants' Aircraft . . . Hearing,* 24–26.

31. U.S. Cong., House, *Merchants' Aircraft . . . Hearing,* 38–39, 44–47.

32. Ibid., 47–54.

33. Ibid., 54–55.

34. U.S. Cong., Senate, *Merchants' Aircraft . . . Hearing,* 24–27.

35. Ibid., 25–32.

36. J. C. Hunsaker, "Outline of Legislative History of McNary-Parker Merchant Airship Bill, Together with Our Contacts with the Aviation Corporation (Del.) and Pan American Airways," 18 February 1931, pp. 3–5, folder 3, box 23, LTA Society Coll., UA. Henry Cord Meyer brought Trippe's opposition to the bill to my attention.

37. Ibid., pp. 3–10.

38. JCH, Merchant Airship Legislation, 1931–32, July 1932, pp. 6–13, folder 4, box 5, LTA Society Coll., UA; *New York Times,* 4 May 1932; U.S. Cong., House, *Congressional Record,* 72d Cong., 1st sess., 75(11,12):12498–504, 13178–91. JCH, Merchant Airship Legislation, 1931–32, p. 12, mistakenly writes that the bill passed on July 16.

39. JCH, Merchant Airship Legislation, 1931–32, July 1932, pp. 16–25; *New York Times,* 18 September 1932.

40. Smith, *Airships* Akron *and* Macon, 31–37.

41. USS *Akron* Court of Inquiry, p. 267, vol. 1, file ZRS 4&5/A17, box 5598, BuAer, Gen. Corresp., 1925–1942, RG 72, NA; see also newspaper clippings in Bird, *Jerome Clarke Hunsaker,* 99–102; Smith, *Airships* Akron *and* Macon, 37–40, 43–44.

42. Smith, *Airships* Akron *and* Macon, 95; *New York Times,* 31 March, 29 April 1931, 25 January 1932; Smith, "Zeppelin Kaput!" 13.

43. *New York Times,* 3 March 1933; U.S. Cong., Senate, *Congressional Record,* 72d Cong., 2d sess. (Washington, D.C.: Government Printing Office, 1933), 76(5):5416; Smith, "Zeppelin Kaput!" 12–13.

44. For accounts of the *Akron*'s last flight, see Smith, *Airships* Akron *and* Macon, 77–82; and Trimble, *Moffett,* 1–3, 264–67.

45. USS *Akron* Court of Inquiry, pp. 266–71, 275–85, vol. 1, file ZRS 4&5/A17, box 5598, BuAer, Gen. Corresp., 1925–1942, RG 72, NA.

46. Finding of Facts, 1 May 1933, USS *Akron* Court of Inquiry, vol. 1, file ZRS 4&5,/A17, box 5598, BuAer, Gen. Corresp., 1925–1942, RG 72, NA.

47. U.S. Cong., *Hearings before a Joint Committee to Investigate Dirigible Disasters,*

73d Cong., 1st sess. (Washington, D.C.: Government Printing Office, 1933), 453– 55, 465–66.

48. Ibid., 478–80.
49. Ibid., 480.
50. Ibid., 482; Trimble, *Moffett,* 133–34, 140.
51. Smith, *Airships* Akron *and* Macon, 91.
52. For a discussion of "lock-in" and "path dependency" by professional engineers, see Timothy R. Whisler, *The British Motor Industry, 1945–1994: A Case Study in Industrial Decline* (Oxford: Oxford University Press, 1999), 8–9, 111, 157.
53. Walter G. Vincenti, "Retractable Airplane Landing Gear and the Northrop 'Anomaly': Variation-Selection and the Shaping of Technology," *Technology and Culture* 35 (January 1994): 2–4, 25.
54. Hunsaker interview, April 1960, OHRO, Columbia University, p. 77. For "internal logic," see Vincenti, *What Engineers Know,* 204.

8. MIT

1. Karl T. Compton to JCH, 31 March 1933, folder 21, box 2, JCHP/MIT.
2. Compton to Paul W. Litchfield, 3 April 1933; Compton to JCH, 11 April 1933; JCH to Compton, 28 April 1933; all in folder 21, box 2, JCHP/MIT; Hunsaker interview by Bonney, 2 November 1971, p. 6, Hunsaker Biography File, NASA History Office.
3. Bush to JCH, 2 May 1933, folder 21, box 2, JCHP/MIT; *New York Times,* 13 June 1933.
4. JCH to Bush, 1 May 1933; JCH to Bush, 19 May 1933; both in folder 21, box 2, JCHP/MIT; *New York Times,* 13 June, 18 August 1933.
5. Alice Hunsaker Bird to author, 1 January 1997; undated clipping from the *Carthage Republican,* in Bird, *Jerome Clarke Hunsaker,* 131; clipping from *Boston Traveler,* 3 March 1947 (courtesy of Alice Hunsaker Bird); grant to Jerome C. Hunsaker, 25 May 1934, Boston (courtesy of Aaron Vidaver, MIT Institute Archives and Special Collections); Jerome C. Hunsaker Jr., interview by Earl A. Thornton, 12 November 1996, Lincoln, Mass.
6. JCH to Reginald Fitz, Thursday Evening Club, 27 November 1946, folder 32, box 1; unidentified clipping, November 1957, folder 41, box 1; both in JCHP/MIT; Alice Hunsaker Bird to author, 12 November 1996.
7. *New York Times,* 15 December 1936, 28 February 1937; see also correspondence in folder 28, box 7, JCHP/NASM; Jerome C. Hunsaker Jr., interview by Earl A. Thornton, 12 November 1996, Lincoln, Mass.; JCH to Philip L. Teed, 16 January 1940, folder 16, box 1, JCHP/MIT.
8. Bird, *Hunsaker Family History,* 13–14; JCH to Gerard Swope, 5 October 1939, folder 15, box 1, JCHP/MIT.
9. F. Robert van der Linden, "Progressives and the Post Office: Walter Folger Brown and the Creation of United States Air Transportation," in William F. Trimble, ed.,

From Airships to Airbus: The History of Civil and Commercial Aviation, vol. 2, *Pioneers and Operations* (Washington, D.C.: Smithsonian Institution Press, 1995), 251–58.

10. *New York Times,* 1, 7, 11 July 1934; telegram, Roosevelt to JCH, 1 July 1934; William B. Stout to JCH, 9 August 1934; JCH to Paul W. Litchfield, 9 August 1934; all in folder 2, box 14, JCHP/NASM.

11. *New York Times,* 11 July 1934; Lucy Fox to E. P. Warner, 17 July 1934; proposed itinerary; both in folder 2, box 14, JCHP/NASM.

12. *New York Times,* 25 September, 3 October 1934.

13. Ibid., 17 October 1934.

14. Ibid., 13 November 1934; Brig. Gen. C. E. Kilbourne to JCH, 15 October 1934, WPD 888-02, file 145.93-97, U.S. Air Force Historical Research Agency, Maxwell Air Force Base, Alabama; see also copies of navy doctrine statements and information on centralized procurement in file 145.93-97, U.S. Air Force Historical Research Agency, Maxwell Air Force Base, Alabama.

15. U.S. Federal Aviation Commission, *Report of the Federal Aviation Commission* (Washington, D.C.: Government Printing Office, 1935), 49–56.

16. Ibid., 243–47; Komons, *Bonfires to Beacons,* 349–51.

17. *Report of the Federal Aviation Commission,* 19–25, 119–20, 123, 127–31.

18. Ibid., 197–206.

19. Hunsaker interview, April 1960, OHRO, Columbia University, p. 36; *Journal of the Aeronautical Sciences* 1 (January 1934): 47–48; *Technology Review* 36 (March 1934): 234.

20. JCH to Hugh Dryden, 26 August 1933, folder 15, box 10, JCHP/NASM; *Journal of the Aeronautical Sciences* 1 (January 1934): i; Levinson, "Encounters between Engineering Science and American Engineering Education," 8; "Biographical Information on Jerome Clarke Hunsaker," p. 11, box 11, Hunsaker Biog. File, Records of NASA, RG 255, NA.

21. JCH, memo. for Finance Committee on Investment Management, 28 February 1941, folder 1, box 21, JCHP/NASM; see also Walter Bonney comments, 2 May 1972, on Hunsaker Papers, NASM, J. C. Hunsaker file, box 2, Misc. Sources, NACA, NASA History Office.

22. Ober, "Story of Aeronautics at MIT," 18–21; Charles C. Bates and John F. Fuller, *America's Weather Warriors, 1814–1985* (College Station: Texas A&M University Press, 1986), 33–34.

23. Ober, "Story of Aeronautics at MIT," 18, 22, 29–30, 33–39, 42; JCH to Clark B. Millikan, 4 December 1933; JCH to Millikan, 26 August 1935; both in Papers of Clark B. Millikan, California Institute of Technology, Institute Archives, Pasadena (courtesy of Richard K. Smith).

24. Ober, "Story of Aeronautics at MIT," 25; JCH to Clark B. Millikan, 4 December 1933, Papers of Clark B. Millikan, Caltech (courtesy of Richard K. Smith).

25. Ober, "Story of Aeronautics at MIT," 16–17; Karl T. Compton, "New Objectives for Technology," *Technology Review* 39 (November 1936): 17.

26. Shatswell Ober, "Development of Aeronautical Engineering at MIT" (Period IV, supplement to "Story of Aeronautics at MIT"), MIT Department of Aeronautics and Astronautics, May 1965, 1–2; *Tech* 57 (26 February 1937): 1, 5; William Heyser, "Technology's Wind Tunnels," *Tech Engineering News* 21 (January 1941): 341, 349.

27. JCH to Charles H. Chatfield, 6 March 1937, folder 8, box 26; JCH to Charles H. Chatfield, 8 July 1938, folder 7, box 26; both in JCHP/NASM; *Tech* 57 (15 October 1937): 1; Ober, "Development of Aeronautical Engineering at MIT," 2–3.

28. J. C. Hunsaker, "The Wright Brothers Wind Tunnel," Institute of the Aeronautical Sciences, 27 January 1939, folder 28, box 2, JCHP/MIT; Ober, "Development of Aeronautical Engineering at MIT," 2–3.

29. Richard K. Smith, "The Intercontinental Airliner and the Essence of Airplane Performance, 1929–1939," *Technology and Culture* 24 (July 1983): 428–49; Richard K. Smith, "The Weight Envelope: An Airplane's Fourth Dimension . . . Aviation's Bottom Line," *Aerospace Historian* 33 (Spring 1986): 30–44; JCH to George J. Mead, 25 March 1936, folder 2, box 27, JCHP/NASM.

30. Richard K. Smith, "The Superiority of the American Transoceanic Airliner, 1932–1939: Sikorsky S-42 vs. Short S. 23," *American Aviation Historical Society Journal* 29 (Summer 1984): 82–94.

31. Jerome C. Hunsaker and George J. Mead, "Around the Corner in Aviation," *Technology Review* 39 (December 1936): 65–68, 80.

32. Rene J. Francillon, *McDonnell Douglas Aircraft since 1920* (Annapolis: Naval Institute Press, 1988), 1:265–68; Peter W. Brooks, *The Modern Airliner: Its Origins and Development* (London: Putnam, 1961), 96.

33. Kenneth Munson and Gordon Swanborough, *Boeing: An Aircraft Album,* no. 4 (New York: Arco, 1972), 71–73.

34. Hunsaker and Mead, "Around the Corner in Aviation," 80–88.

35. *New York Times,* 20 September, 10 October 1936; Richard K. Smith, "Notes on the International Zeppelin Transport Co. (IZT) and Pacific Zeppelin Transport Co. (PZT)," *Buoyant Flight* 39 (July–August 1992): 4–5. Henry Cord Meyer brought the 1936 *Hindenburg* flight to my attention.

36. JCH to G. W. Raehmel, 22 September 1937; cable, M. L. Ham to Col. H. Walaardt Sacre, in Holland, no date; M. L. Ham to Col. H. Walaardt Sacre, 1 October 1937; JCH to Capt. J. A. Furer, 31 August 1937; all in folder 10, box 2, JCHP/MIT; *New York Times,* 29 September 1937.

37. J. C. Hunsaker, "The Development of Trans-Atlantic Aircraft," 1937, folder 99, box 3, JCHP/MIT.

38. JCH to Hugo Eckener, 31 August 1937; JCH to Capt. J. A. Furer, 31 August 1937; JCH to President Adolf Baeumker, Lilienthal Gesellschaft, 3 November 1937; D. R. Pye to JCH, 8 September 1937; all in folder 10, box 2, JCHP/MIT; *New York Times,* 2 November 1937.

39. JCH to George J. Mead, 12 July 1937, folder 8, box 26, JCHP/NASM.

40. George J. Mead to JCH, 13 January 1938, folder 7, box 26, JCHP/NASM.

41. Hunsaker interview, April 1960, OHRO, Columbia University, pp. 82–83.
42. Smith, "Intercontinental Airliner," 446; J. E. Slater to JCH, 3 January 1939, folder 16, box 10, JCHP/NASM; see also R. E. G. Davies, *Airlines of the United States since 1914* (London: Putnam, 1972), 359–60.
43. JCH to J. E. Slater, 10 May 1939; JCH to Slater, 2 June 1939; JCH to Slater, 31 July 1939; all in folder 16, box 10, JCHP/NASM.
44. JCH to Slater, 2 June 1939; JCH to C. L. Egtvedt, 2 June 1939; both in folder 16, box 10, JCHP/NASM.
45. JCH to Slater, 31 July 1939; Slater to JCH, 10 August 1939; both in folder 16, box 10, JCHP/NASM; Davies, *Airlines of the United States since 1914,* 361–62.
46. D. G. Richardson to JCH, 28 March 1941; JCH to D. G. Richardson, 9 April 1941; both in folder 1, box 11, JCHP/NASM.
47. JCH to D. G. Richardson, 24 June 1941, folder 1, box 11, JCHP/NASM.
48. Eugene E. Wilson, *Slipstream: The Autobiography of an Air Craftsman,* 3d ed. (Palm Beach, Fla.: Literary Investment Guild, 1967), 164–67, 170.
49. J. C. Hunsaker, "Report on Propeller Development," 7 July 1937, folder 8, box 26, JCHP/NASM.
50. Ibid.
51. Memo, Frank Caldwell to George J. Mead, 3 August 1937, folder 8, box 26, JCHP/NASM.
52. JCH to Charles H. Chatfield, 8 August 1940, folder 6, box 26, JCHP/NASM.
53. Preliminary Study, Proposed United Aircraft Wind Tunnel, 25 February 1941; JCH to John G. Lee, 12 March 1941; John G. Lee to Frank Caldwell, 26 May 1941; E. E. Wilson to JCH, 11 July 1941; all in folder 5, box 26, JCHP/NASM.

9. War and the NACA

1. Itinerary, folder 4, box 4, JCHP/MIT; *Opelika* (Ala.) *Daily News,* 8 May 1939.
2. JCH to George Lewis, 1 September 1926, Hunsaker, J. C., 1916–35 (1) (M-N) folder, box 12, Hunsaker Biog. File, Records of NASA, RG 255, NA; G. W. Lewis to JCH, 23 August 1938, Misc. Sources, NACA, NASA History Office.
3. Roland, *Model Research,* 2:439, 448, 450; Hansen, *Engineer in Charge,* 5–6.
4. Roland, *Model Research,* 1:149–50.
5. JCH, "Research Bearing on the Future of Aviation," MIT, 29 March 1939, folder 7, box 4, JCHP/MIT.
6. Roland, *Model Research,* 1:168–69.
7. JCH to Philip L. Teed, 16 January 1940, folder 16, box 1, JCHP/MIT.
8. Roland, *Model Research,* 1:158–60; Elizabeth A. Muenger, *Searching the Horizon: A History of Ames Research Center, 1940–1976* (Washington, D.C.: National Aeronautics and Space Administration, 1985), 1–7; Lindbergh to JCH, 4 September 1939, C. A. Lindbergh file, Misc. Sources, NACA, NASA History Office.

9. Roland, *Model Research,* 1:160–66; Virginia P. Dawson, *Engines and Innovation: Lewis Laboratory and American Propulsion Technology* (Washington, D.C.: National Aeronautics and Space Administration, 1991), 6–15.

10. Edward Warner to Vannevar Bush, 27 December 1940; JCH to Bush, 24 January 1941; both in NACA Gen. Corresp. folder, box 10, Hunsaker Biog. File, Records of NASA, RG 255, NA.

11. Julius Augustus Furer, *Administration of the Navy Department in World War II* (Washington, D.C.: U.S. Navy Department, Naval History Division, 1959), 775–76; Hunsaker interview, April 1960, OHRO, Columbia University, p. 100; memo, JCH to SecNav, 7 April 1941, folder 27, box 1, JCHP/MIT.

12. Furer, *Administration of the Navy Department in World War II,* 776–80.

13. "The Evolution of the Office of Naval Research," Research and Development Reports File, box 7, Papers of Julius A. Furer, Library of Congress Manuscript Division (hereafter cited as Furer Papers, LCMD); 18 March–11 July 1941 entries, JCH Diary, 1941–1942, folder 11, box 2, JCH/MIT; Hunsaker interview, April 1960, OHRO, Columbia University, p. 98.

14. Furer, *Administration of the Navy Department in World War II,* 782; "The Evolution of the Office of Naval Research," Research and Development Reports File; JCH to Furer, 19 July 1945; both in box 7, Furer Papers, LCMD; Vannevar Bush to members of advisory board, OSRD, 19 December 1941, Hunsaker, J. C., 1938–1944 (2) folder, box 11, Hunsaker Biog. File, Records of NASA, RG 255, NA.

15. Roland, *Model Research,* 2:489; JCH memo, Suggestions for Changes in Organization and Procedure, 1941, NACA Gen. Corresp. folder, box 10, Hunsaker Biog. File, Records of NASA, RG 255, NA.

16. Roland, *Model Research,* 1:169–70; G. Pascal Zachary, *Endless Frontier: Vannevar Bush, Engineer of the American Century* (Cambridge: MIT Press, 1999), 129–30, 139; Bush to JCH, 19 June 1963, folder 1, box 2, JCHP/MIT.

17. Minutes of regular meeting of Executive Committee, 24 July 1941; minutes of annual meeting of NACA, 23 October 1941; both in Misc. Sources, NACA, NASA History Office.

18. U.S. Cong., House, *Hearings before the Subcommittee of the Committee on Appropriations on Second Supplemental National Defense Appropriations Bill for 1942,* 77th Cong., 1st sess. (Washington, D.C.: Government Printing Office, 1941), 337–44.

19. Roland, *Model Research,* 1:173–76; Muenger, *Searching the Horizon,* 10–17; JCH to Mead, 28 November 1941, NACA Gen. Corresp. folder, box 10, Hunsaker Biog. File, Records of NASA, RG 255, NA.

20. 7, 8 December 1941 entries, JCH Diary, 1941–42, folder 11, box 2, JCHP/MIT.

21. U.S. Cong., Senate, *Hearings before the Subcommittee of the Committee on Appropriations, Third Supplemental National Defense Appropriation Bill for 1942,* 77th Cong., 1st sess. (Washington, D.C.: Government Printing Office, 1941), 196–202.

22. U.S. National Advisory Committee for Aeronautics, *Twenty-Seventh Annual Report of the National Advisory Committee for Aeronautics, 1941* (Washington, D.C.: Government Printing Office, 1942), 1–3.
23. Ibid., 8; Wilson to JCH, 3, 14 October 1941, NACA Gen. Corresp. folder, box 10, Hunsaker Biog. File, Records of NASA, RG 255, NA.
24. Roland, *Model Research,* 1:182–83; Hansen, *Engineer in Charge,* 212.
25. JCH to Lindbergh, 6 August 1942, file 101.1, Jerome C. Hunsaker, Office of Scientific Research and Development folder, box 12, Hunsaker Biog. File, Records of NASA, RG 255, NA; 7–21 August 1942 entries, JCH Diary, 1941–42, folder 11, box 2; JCH to Warren Weaver, 17 February 1945, folder 24, box 4; JCH to Capt. David Swope, 17 July 1944, folder 32, box 1; all in JCHP/MIT.
26. J. C. Hunsaker, *Forty Years of Aeronautical Research,* from *Smithsonian Report for 1955* (Washington, D.C.: Smithsonian Institution, 1956), 266–67; Dawson, *Engines and Innovation,* 6; JCH to Mead, 18 October 1941, NACA Gen. Corresp. folder, box 10, Hunsaker Biog. File, Records of NASA, RG 255, NA; Mead to JCH, 22 December 1942, Misc. Sources, NACA, NASA History Office; Hunsaker interview, April 1960, OHRO, Columbia University, pp. 104–5.
27. JCH, "Report on Aeronautics," National Academy of Sciences, Washington, D.C., 22 November 1943, folder 29, box 1, JCHP/MIT.
28. Roland, *Model Research,* 1:189; Hansen, *Engineer in Charge,* 230–33; Dawson, *Engines and Innovation,* 60; JCH to A. G. Herreshoff, 15 April 1941, Hunsaker Biography File, NASA History Office.
29. JCH to E. E. Wilson, 30 June 1941; L. S. Hobbs to JCH, 12 December 1941; JCH to Hobbs, 18 December 1941; all in folder 5, box 26, JCHP/NASM.
30. Roland, *Model Research,* 1:191; Edward W. Constant II, *The Origins of the Turbojet Revolution* (Baltimore: Johns Hopkins University Press, 1980), 222–23.
31. JCH to Paul E. Klopstag, 29 January 1945, NACA Gen. Corresp. 1945 folder, box 14, Hunsaker Biog. File, Records of NASA, RG 255, NA.
32. L. S. Hobbs to JCH, 29 September 1941; JCH to L. S. Hobbs, 2 October 1941; both in box 10, Hunsaker Biog. File, Records of NASA, RG 255, NA; Mead to JCH, 22 December 1942, Misc. Sources, NACA, NASA History Office; Roland, *Model Research,* 1:191–92.
33. Dawson, *Engines and Innovation,* 41–42; Hansen, *Engineer in Charge,* 247; Roland, *Model Research,* 1:193–94; JCH to Bush, 26 April 1944, A-19, Personal File, Dr. Hunsaker, Correspondence and Report File, 1942-60, box 207, Records of NASA, RG 255, NA.
34. Report of Conference on Jet Propulsion in England, 18 December 1943, Hunsaker Papers 1940s folder, box 13, Hunsaker Biog. File, Records of NASA, RG 255, NA.
35. Theodore von Kármán, with Lee Edson, *The Wind and Beyond: Theodore Von Kármán, Pioneer in Aviation and Pathfinder in Space* (Boston: Little, Brown, 1967), 243.
36. JCH to A. G. Herreshoff, 15 April 1941, Hunsaker Biography File, NASA History

Office; JCH, "Report on Aeronautics," National Academy of Sciences, 22 November 1943, pp. 6, 12, folder 29, box 1, JCHP/MIT; Joseph Adams Shortal, *A New Dimension: Wallops Island Flight Test Range, the First Fifteen Years* (Washington, D.C.: National Aeronautics and Space Administration, 1978), 16, 21.

37. U.S. National Advisory Committee for Aeronautics, *Twenty-Seventh Annual Report . . . 1941,* 4; JCH, "Aeronautical Research," American Academy of Arts and Sciences, 9 February 1944, folder 30, box 1, JCHP/MIT.

38. JCH, "Aeronautical Research," American Academy of Arts and Sciences, 9 February 1944, folder 30, box 1, JCHP/MIT.

39. JCH Diary, 1941–1942, folder 11, box 2, JCHP/MIT; Hunsaker, Jerome C., 1941–1944 and 1945 folders, 31-1 file, box 2, Records Relating to NACA Committees and Subcommittees (Decimal File), 1918–1951; JCH to Lewis Strauss, 21 March 1956, NACA, January to June 1956 folder, box 17, Hunsaker Biog. File; both in Records of NASA, RG 255, NA.

40. Roland, *Model Research,* 2:472–73, 487, 489.

41. *Technology Review* 62 (November 1959): 85, 112; JCH Diary, 1941–1942, folder 11, box 2, JCHP/MIT; news release, 14 April 1941, Hunsaker, Jerome C., file, Operational Archives Branch, Naval Historical Center; 15 November 1942 entry, JCH Diary, 1941–1942, folder 11, box 2, JCHP/MIT; JCH to George Mead, 25 March 1943, Hunsaker, J.C., 1938–1944(2) (P-Educ.) Folder, box 11, Hunsaker Biog. File, Records of NASA, RG 255, NA.

42. Joseph C. Boyce, ed., *New Weapons for Air Warfare: Fire- Control Equipment, Proximity Fuzes, and Guided Missiles,* in *Science in World War II, Office of Scientific Research and Development* (Boston: Little, Brown, 1947), 249–51.

43. Ibid., 251–52; L. O. Grondahl, "The High Angle Dirigible Bomb Project," 24 November 1942, folder 6, box 23; L. O. Grondahl to H. B. Richmond, 2 December 1942, folder 6, box 23; JCH to Grondahl, 9 February 1943, folder 3, box 23; all in JCHP/NASM; JCH to Grondahl, 14 October 1947, folder 18, box 1, JCHP/MIT.

44. JCH to Ascher H. Shapiro, 3 June 1966, folder 7, box 16; JCH to Charles Platrier, 21 November 1944, folder 4, box 17; reprint from *Nature,* 7 June 1947, folder 2, box 17; all in JCHP/NASM.

45. Minutes of Regular meeting of Executive Committee, 27 January 1944, Minutes, January–June 1944 folder, p. 10, Minutes of the Executive Committee of the NACA, 1943–1945, box 9, Records of NASA, RG 255, NA; JCH, "Memorandum on Postwar Research Policy for NACA," 27 July 1944, in Roland, *Model Research,* 2:684–85.

46. "Notes on Discussion at Meeting of NACA," 8 August 1944, in Roland, *Model Research,* 2:686–90.

47. JCH Memorandum, "Guided Missiles, NACA Program and Facilities," 15 December 1944; JCH to Bush, 16 December 1944; both in 101.1 Jerome C. Hunsaker, Office of Scientific Research and Development folder, box 12, Hunsaker Biog. File, Records of NASA, RG 255, NA.

48. U.S. Cong., House, *Hearings before the Select Committee on Post-War Military Policy,* 78th Cong., 2d sess. (Washington, D.C.: Government Printing Office, 1945), 259–61.

49. Ibid., 262–66, 269, 271.

50. Minutes of Regular meeting of Executive Committee, 18 December 1944, p. 2, Minutes, July–December 1944 folder, box 9, Minutes of the Executive Committee of the NACA, 1943–1945, Records of NASA, RG 255, NA.

10. New Directions and Old

1. Roland, *Model Research,* 1:203–4; U.S. Cong., Senate, *Hearings before a Special Committee Investigating the National Defense Program,* 79th Cong., 2d sess. (Washington, D.C.: Government Printing Office, 1946), 16807.

2. U.S. Cong., Senate, *Hearings before a Special Committee Investigating the National Defense Program,* 16808, 16810–13, 16815, 16843–46.

3. Ibid., 16817–21, 16853.

4. Ibid., 16828–33, 16842.

5. J. C. Hunsaker, "Jet Propulsion," National Academy of Sciences, 23 April 1946, pp. 2–4, 7–9, Hunsaker Biography File, NASA History Office.

6. Zachary, *Endless Frontier,* 316–17.

7. Roland, *Model Research,* 1:207; Minutes of Annual Meeting and Minutes of Special (Organization) Meeting of Executive Committee, 25 October 1945, p. 3, Minutes, July–December 1945 folder, box 9, Minutes of the Executive Committee of the NACA, 1943–1945, Records of NASA, RG 255, NA; Vannevar Bush to JCH, 9 December 1949, NACA Gen. Corresp. 1946 folder, box 14, Hunsaker Biog. File, Records of NASA, RG 255, NA.

8. JCH to Bush, 24 December 1946, NACA Gen. Corresp. 1946 folder, box 14, Hunsaker Biog. File, Records of NASA, RG 255, NA. Portions of this letter appear in Roland, *Model Research,* 1:209.

9. JCH to H. M. Horner, 19 November 1948, NACA Gen. Corresp. July–December 1948 folder, box 15, Hunsaker Biog. File, Records of NASA, RG 255, NA.

10. Memorandum: Guided Missiles, NACA Program and Facilities, 15 December 1944, NACA Gen. Corresp. 1944 folder, box 14, Hunsaker Biog. File, Records of NASA, RG 255, NA.

11. U.S. Cong., Senate, *Hearings before a Special Committee Investigating the National Defense Program,* 16833, 16846–49, 16852.

12. Ibid., 16823–24; Roland, *Model Research,* 1:211–14; U.S. Cong., Senate, *Hearings before a Special Committee Investigating the National Defense Program,* 16832; Hugh Dryden to JCH, 15 June 1953, A-19, Personal File, Dr. Hunsaker, 1953, Correspondence and Report File, 1942–1960, box 207, Records of NASA, RG 255, NA.

13. U.S. Cong., House, *Hearings before the Subcommittee of the Committee on Appropriations on the Independent Offices Appropriation Bill for 1950,* 81st Cong.,

1st sess. (Washington, D.C.: Government Printing Office, 1949), 894–95. The Civil Aeronautics Administration and the Civil Aeronautics Board (CAB) succeeded the Civil Aeronautics Authority in 1940.

14. Roland, *Model Research,* 1:214–16.

15. Ibid., 1:215–16; JCH to Raymond Committee members, 26 April 1946, NACA Gen. Corresp. 1946 folder, box 14, Hunsaker Biog. File; Minutes of Regular Meeting of Executive Committee, 6 June 1946, pp. 4–6, Minutes, January–June 1946 folder, box 10, Minutes of the Executive Committee of the NACA, 1946–1948; NACA, A Proposal for the Construction of a National Supersonic Research Center, April 1946, MLR, 1A, file 521–32, box 178, Gen. Corresp. (Decimal) 1929–1952 (courtesy of J. Lawrence Lee); all in Records of NASA, RG 255, NA; George Lewis to JCH, 26 June and 3, 10, 16 July 1946, folder 35, box 2, JCHP/MIT.

16. JCH diary, June 20–July 26, 1946; John Victory to J. J. Ide, 7 June 1946; JCH to Ide, 12 June 1946; JCH to George Lewis, 6 July 1946; all in folder 34, box 2, JCHP/MIT.

17. "Newton and Fluid Mechanics," Remarks by Dr. J. C. Hunsaker at Newton Tercentenary Celebration, England, 1946, J. C. Hunsaker file, box 2, Misc. Sources, NACA, NASA History Office; for a general discussion of the transonic problem and the NACA's approaches to it, see Hansen, *Engineer in Charge,* 249–70.

18. Memorandum re travel: Visit to Europe, 1946, 31 July 1946, Hunsaker, J. C., 1945–1946 (2)(P-Educ.), box 11, Hunsaker Biog. File, Records of NASA, RG 255, NA; JCH diary, 20 June–26 July 1946, folder 34, box 2; JCH to George Lewis, 6, 14, 21 July 1946, folder 34, box 2; JCH, Memorandum for NACA Files: "Visit to de Havilland," August 1946, folder 35, box 2; JCH, Memorandum for NACA Files: "Rolls Royce Gas Turbines," August 1946, folder 35, box 2; JCH to J. J. Ide, 31 July 1946, folder 32, box 1; all in JCHP/MIT.

19. JCH to Ide, 4 October 1945, J. J. Ide and Paris Office file, J. C. Hunsaker file, box 2, Misc. Sources, NACA, NASA History Office; JCH to George Lewis, 6, 14 July 1946, folder 34, box 2, JCHP/MIT; Roland, *Model Research,* 1:241.

20. Minutes of Meeting of Special Committee on Supersonic Facilities, 16 August 1946, pp. 2–6; Minutes of Meeting of Special Committee on Supersonic Facilities, 21, 22, 24 October 1946, pp. 5–6, 9; both in NACA Gen. Corresp. 1946 folder, box 14, Hunsaker Biog. File, Records of NASA, RG 255, NA.

21. Minutes of Meeting of Special Committee on Supersonic Facilities, 21, 22, 24 October 1946, pp. 18–20, NACA Gen. Corresp. 1946 folder, box 14, Hunsaker Biog. File; Some Notes on Discussion of Report of Special Committee on Supersonic Facilities, Annual Meeting, NACA, 24 October 1946, Minutes July–December 1946, box 10, Minutes of the Executive Committee of the NACA, 1946–1948; both in Records of NASA, RG 255, NA.

22. Some Notes on Discussion of Report of Special Committee on Supersonic Facilities, Annual Meeting, NACA, 24 October 1946, Minutes July–December 1946, box 10, Minutes of the Executive Committee of the NACA, 1946–1948, Records of NASA, RG 255, NA.

23. Minutes of Meeting of Special Committee on Supersonic Facilities, 19 December 1946, pp. 3, 5–9, NACA Gen. Corresp. 1946 folder, box 14, Hunsaker Biog. File; Hugh L. Dryden to JCH, 7 July 1949, NACA Corresp. July–December 1949 folder, box 15; both in Records of NASA, RG 255, NA.

24. JCH to George Lewis, 26 December 1946, NACA Gen. Corresp. 1946 folder, box 14, Hunsaker Biog. File, Records of NASA, RG 255, NA; Richard P. Hallion, *On the Frontier: Flight Research at Dryden, 1946–1981* (Washington, D.C.: National Aeronautics and Space Administration, 1984), 11–27, 34, 39.

25. James R. Hansen, "George W. Lewis and the Management of Aeronautical Research," in William M. Leary, ed., *Aviation's Golden Age: Portraits from the 1920s and 1930s* (Iowa City: University of Iowa Press, 1989), 93–112; Hunsaker interview by Bonney, 2 November 1971, p. 11, Hunsaker Biography File, NASA History Office; JCH to Lewis, 28 June 1948, NACA Gen. Corresp. January–June 1948 folder, box 15, Hunsaker Biog. File, Records of NASA, RG 255, NA. The Lewis laboratory was renamed the John H. Glenn Research Center at Lewis Field on 1 March 1999 to honor the Ohio astronaut and senator.

26. JCH to Dryden, 26 August 1933, folder 15, box 10, JCHP/NASM; JCH to Dryden, 30 June 1947, NACA Gen. Corresp. January–June 1947 folder, box 14, Hunsaker Biog. File, Records of NASA, RG 255, NA; Hansen, *Engineer in Charge,* 217; Roland, *Model Research,* 1:226.

27. JCH to George Lewis, 8 May 1947, NACA Gen. Corresp. January–June 1947 folder, box 14, Hunsaker Biog. File, Records of NASA, RG 255, NA; Hansen, *Engineer in Charge,* 343–47.

28. JCH to George Lewis, 8 May 1947, NACA Gen. Corresp. January–June 1947 folder, box 14, Hunsaker Biog. File, Records of NASA, RG 255, NA; Jingsheng Dong, "Biography of Hsue-shen Tsien: The Story of an American-Trained Chinese Rocket Expert" (Ph.D. diss., Auburn University, 1996), 47–57, 123–29.

29. JCH to Vannevar Bush, 26 September 1947, NACA Gen. Corresp. July–December 1947 folder, Hunsaker Biog. File, box 15, Records of NASA, RG 255, NA; U.S. President's Air Policy Commission, Unclassified Testimony before the President's Air Policy Commission, 3:1–9, 850–66; *Survival in the Air Age: A Report by the President's Air Policy Commission* (Washington, D.C.: Government Printing Office, 1948), 49–65, 76–85.

30. George Lewis to JCH, 16 February 1948, NACA Gen. Corresp. January–June 1948 folder, box 15, Hunsaker Biog. File, Records of NASA, RG 255, NA; JCH to R. V. Southwell, 6 April 1948, folder 2, box 17, JCHP/NASM; JCH to Elmer B. Staats, Asst. Director, Bureau of the Budget, 10 March 1948, NACA Gen. Corresp. January–June 1948 folder, box 15, Hunsaker Biog. File, Records of NASA, RG 255, NA.

31. Roland, *Model Research,* 1:283–84, 2:435; Dryden to JCH, 7 July 1949, NACA Gen. Corresp. July–December 1949 folder, box 15; JCH to Doolittle, 12 July 1949, NACA Gen. Corresp. July–December 1949 folder, box 15; Dryden, memo for NACA Panel on Facilities, 23 November 1949, NACA Gen. Corresp. January–

June 1949 folder, box 16; all in Hunsaker Biog. File, Records of NASA, RG 255, NA.

32. U.S. Cong., House, *Hearings before the Subcommittee of the Committee on Appropriations on the Independent Offices Appropriation Bill for 1951,* 81st Cong., 2d sess. (Washington, D.C.: Government Printing Office, 1950), 383–85. Information on Unitary Plan tunnels courtesy of J. Lawrence Lee.

33. Jeffrey G. Barlow, *Revolt of the Admirals: The Fight for Naval Aviation, 1945–1950* (Washington, D.C.: Naval Historical Center, 1994), 174–91, 215–83.

34. JCH to A. W. Kimbell, 22 November 1949, folder 34, box 1; JCH notes for seminar, 5 October 1951, folder 52, box 4; both in JCHP/MIT; interview with Dr. Jerome C. Hunsaker, 27 March 1962, p. 3, Hunsaker Biography File, NASA History Office.

35. JCH to B. E. Hutchinson, 17 July 1950, folder 35, box 1; JCH notes, "The Problems of Naval Aviation," 13 December 1954, folder 38, box 1; both in JCHP/MIT.

36. Bernard J. Snyder, *Aircraft Nuclear Propulsion: An Annotated Bibliography* (Washington, D.C.: USAF History and Museums Program, 1996), v; JCH, "Jet Propulsion," National Academy of Sciences, 23 April 1946, Hunsaker Biography File, NASA History Office; Lewis L. Strauss to JCH, 24 October 1947, folder 1, box 13, JCHP/NASM.

37. Minutes of Regular Meeting of Executive Committee, 13 September 1945, p. 8, Minutes, July–December 1945 folder, box 9, Minutes of the Executive Committee of the NACA, 1943–1945, Records of NASA, RG 255, NA; Dawson, *Engines and Innovation,* 77–78; E. U. Condon to JCH, 16 April 1947, in Snyder, *Aircraft Nuclear Propulsion,* 257–58.

38. Lewis L. Strauss to JCH, 27 October 1954, NACA, 1 July–31 December 1954 folder, box 17, Hunsaker Biog. File, Records of NASA, RG 255, NA; *New York Times,* 25 September 1955; Dawson, *Engines and Innovation,* 184.

39. JCH, "NACA 1915–1955," draft 8-2-55, p. 63, J. C. Hunsaker file, box 2, Misc. Sources, NACA, NASA History Office (published in 1956 as *Forty Years of Aeronautical Research*); clipping, *New York Herald Tribune,* 23 January 1956, in Hunsaker, J. C., 1952– (2)(P-Educ.) folder, box 11, Hunsaker Biog. File, Records of NASA, RG 255, NA.

40. First meeting of the Ad Hoc Committee on ANP Hazards, 12 November 1957; second meeting of the Ad Hoc Committee on ANP Hazards, 9–10 December 1957; both in folder 1, box 10, JCHP/NASM.

41. John R. M. Wilson, *Turbulence Aloft: The Civil Aeronautics Administration Amid Wars and Rumors of Wars, 1938–1953* (Washington, D.C.: U.S. Department of Transportation, Federal Aviation Administration, 1979), 249–56; Truman to JCH, 15 June 1947, folder 9, box 23, JCHP/NASM; minutes of Regular Meeting of Executive Committee, 15 January 1948, p. 7, Minutes, January–June 1948 folder, box 10, Minutes of the Executive Committee of the NACA, 1946–1948, Records of NASA, RG 255, NA.

42. Wilson, *Turbulence Aloft,* 259–62; *The Airport and Its Neighbors: The Report of*

the President's Airport Commission (Washington, D.C.: Government Printing Office, 1952), iii–vii, 101–2.

43. *The Airport and Its Neighbors,* 10–12, 14–15, 16–21; Wilson, *Turbulence Aloft,* 262–64.

44. D. S. Fahrney, "History of Pilotless Aircraft and Guided Missiles," undated ms., probably 1958, Operational Archives Branch, Naval Historical Center, Washington, D.C., 1270–73.

45. Ibid., p. 1272; Ober, "Story of Aeronautics at MIT," 6–8; JCH to George Lewis, 8 May 1947, NACA Gen. Corresp. January–June 1947 folder, box 14, Hunsaker Biog. File; Hugh Dryden to JCH, 15 June 1953, A-19, Personal File, Dr. Hunsaker, 1953, Correspondence and Report File, 1942–1960, box 207; both in Records of NASA, RG 255, NA; *New York Times,* 15 June 1947, 2 December 1949.

46. A. N. Richards to members of NAS Council, 10 March 1948; telegram, JCH to Richards, 12 March 1948; Richards to members of NAS Council, 7 April 1948; JCH to Richards, 12 April 1948; all in folder 5, box 20, JCHP/NASM.

47. JCH to Edward P. Farley, 31 March 1948, folder 33, box 1, JCHP/MIT. See also Walter Bonney comments, 2 May 1972, on box 5 of the Hunsaker Papers, NASM, J. C. Hunsaker file, Misc. Sources, NACA, NASA History Office.

48. JCH to Edwin B. Goodell Jr., 1 November 1951, folder 22, box 2, JCHP/MIT; additional information on the Struik affair in folder 36, box 1, JCHP/MIT. See also David Caute, *The Great Fear: The Anti-Communist Purge under Truman and Eisenhower* (New York: Simon and Schuster, 1978), 410–11.

49. JCH to Hugh L. Dryden, 30 January 1948; JCH to J. F. Victory, 12 August 1948; both in folder 2, box 17, JCHP/NASM; *New York Times,* 20 August 1948.

50. JCH to John F. Victory, 2 July 1952; JCH, "Social Aspects of Aeronautics," 21 August 1952; both in A-19, Personal File, Dr. Hunsaker, Correspondence and Report File, 1942–1960, box 207, Records of NASA, RG 255, NA; JCH to Sir David Pye, 27 May 1952, folder 13, box 2, JCHP/MIT.

51. Sir Harry M. Garner to JCH, 9 September 1952, folder 13, box 2; George R. Harrison to JCH, 10 November 1952, folder 13, box 2; JCH to Prof. J. Kampe' de Ferier, 11 March 1957, folder 41, box 1; all in JCHP/MIT; JCH to Prof. Kerim Erim, 1 July 1952, folder 10, box 16, JCHP/NASM; *New York Times,* 26 September 1952.

52. Minutes of Regular Meeting of Executive Committee, 21 July 1949, p. 11, Minutes, July–December 1949 folder, box 11, Minutes of the Executive Committee of the NACA, 1949–1952, Records of NASA, RG 255, NA; JCH to Sir George Leigh-Jones, 15 February 1956, folder 40, box 1, JCHP/MIT; Alice Hunsaker Bird to author, 2 December 1996; Jerome C. Hunsaker Jr., interview by Earl A. Thornton, 12 November 1996, Lincoln, Mass.

53. Roland, *Model Research,* 2:473; U.S. Cong., House, *Hearings . . . on the Independent Offices Appropriation Bill for 1950,* 881; U.S. Cong., House, *Hearings before the Subcommittee of the Committee on Appropriations on the Independent Offices Appropriation Bill for 1949,* 80th Cong., 2d sess. (Washington, D.C.: Government Printing Office, 1948), 520–22.

54. U.S. Cong., House, *Hearings . . . on the Independent Offices Appropriation Bill for 1949,* 523–25, 536–38; Roland, *Model Research,* 2:473, 475.

55. U.S. Cong., House, *Hearings . . . on the Independent Offices Appropriation Bill for 1950,* 864–65, 874–80.

56. Ibid., 882–84, 890–91; U.S. Cong., House, *Hearings . . . on the Independent Offices Appropriation Bill for 1951,* 391–92; Roland, *Model Research,* 2:473. A \$75 million deficiency appropriation for fiscal year 1950 was earmarked for the NACA's Unitary Plan wind tunnel program.

57. U.S. Cong., House, *Hearings . . . on the Independent Offices Appropriation Bill for 1951,* 370, 375, 394, 429; U.S. Cong., House, *Hearings before the Subcommittee of the Committee on Appropriations on the Independent Offices Appropriation Bill for 1952,* 82d Cong., 1st sess. (Washington, D.C.: Government Printing Office, 1951), pt. 1, 346–48; Roland, *Model Research,* 2:473–75.

58. Telegram, JCH to Sen. Leverett Saltonstall, 11 August 1950, NACA Gen. Corresp. July–December 1950 folder, Hunsaker Biog. File, box 16, Records of NASA, RG 255, NA.

59. U.S. Cong., House, *Hearings . . . on the Independent Offices Appropriation Bill for 1952,* pt. 1, 346–60, 362–63, 365, 367, 375–80.

60. Roland, *Model Research,* 1:264–67, 2:474–75; U.S. Cong., House, *Hearings . . . on the Independent Offices Appropriation Bill for 1952,* pt. 1, 385; U.S. Cong., House, *Hearings before the Subcommittee of the Committee on Appropriations on the Independent Offices Appropriation Bill for 1953,* 82d Cong., 2d sess. (Washington, D.C.: Government Printing Office, 1952), pt. 1, 342–45.

61. U.S. Cong., House, *Hearings . . . on the Independent Offices Appropriation Bill for 1953,* pt. 1, 345–46, 361–62, 367–69, 388–89, 399, 403–5; Roland, *Model Research,* 2:474–75.

11. The Long Twilight

1. *Tech* 71 (6 April 1951): 1, 8; J. R. Markham, statement at retirement banquet for JCH, 24 May 1952, folder 27, box 2, JCHP/MIT; Ober, "Story of Aeronautics at MIT," 27.

2. Ober, "Story of Aeronautics at MIT," 27.

3. Memo, JCH to J. R. Killian, 16 August 1948, folder 1, box 16; JCH to J. R. Killian, 15 February 1952, folder 22, box 2; both in JCHP/NASM.

4. JCH, "Aeronautical Engineering Department, MIT, 1933–1960," pp. 11–12, folder 21, box 2, JCHP/MIT.

5. Ibid., pp. 8–9; Ober, "Story of Aeronautics at MIT," 24; Edward Warner to JCH, 18 June 1954, folder 38, box 1, JCHP/MIT.

6. Killian to JCH, 13 December 1965, folder 25, box 2, JCHP/MIT. The formal agreement for the Maclaurin Professorship is in this folder. See also Bird, *Jerome Clarke Hunsaker,* 188–89.

7. See various correspondence in National Academy of Sciences, folders 1–4, box 21, JCHP/NASM.

8. Hunsaker, Jerome C. file, Operational Archives Branch, Naval Historical Center; JCH to Hugh L. Dryden, 11 February 1955; Capt. T. H. Moorer to JCH, 23 August 1955; Schedule for European Trip, 3–14 September 1955; all in folder 3, box 24, JCHP/NASM.

9. JCH to Leverett Saltonstall, 6 April 1954, NACA, January 1–June 30, 1954 folder, box 17, Hunsaker Biog. File, Records of NASA, RG 255, NA; Roland, *Model Research,* 1:267–71, 277–79, 2:474–75; U.S. National Advisory Committee for Aeronautics, *Fortieth Annual Report of the National Advisory Committee for Aeronautics, 1954* (Washington, D.C.: Government Printing Office, 1956), 74.

10. JCH to Eisenhower, 6 August 1954, NACA, 1 July–31 December 1954 folder, box 17, Hunsaker Biog. File, Records of NASA, RG 255, NA; Roland, *Model Research,* 1:279–80; U.S. National Advisory Committee for Aeronautics, *Forty First Annual Report of the National Advisory Committee for Aeronautics, 1955* (Washington, D.C.: Government Printing Office, 1956), 68.

11. Ludwell Lee Montague, *General Walter Bedell Smith as Director of Central Intelligence* (University Park: Pennsylvania State University Press, 1992), 175–76; JCH, "Aeronautical Engineering Department, MIT, 1933–1960," p. 4, folder 21, box 2; JCH to George Hewitt Myers, 24 February 1953, folder 37, box 1; Harold O. Jenkins, CIA, to JCH, 31 August 1955, folder 5, box 3; JCH to Cornelius Van S. Roosevelt, 17 August 1955, folder 5, box 3; all in JCHP/MIT. A 1999 Freedom of Information Act request to examine CIA records revealed no further information on these committees.

12. JCH to Hugh L. Dryden, 8 May 1956; Dryden to JCH, 10 May 1956; both in folder 40, box 1, JCHP/MIT; Jay Miller, *Lockheed U-2* (Austin, Texas: Aerofax, 1983), 12–13, 19–25; Ben R. Rich and Leo Janos, *Skunk Works: A Personal Memoir of My Years at Lockheed* (Boston: Little, Brown, 1994), 145.

13. For a thorough critique of Majestic 12 documents, see Philip J. Klass, "Special Report: The MJ-12 Crashed-Saucer Documents," *Skeptical Inquirer* 12 (Winter 1987–88): 137–46. For copies of the documents, see websites set up by UFO researchers. One such is http://site034145.primehost.com/ds/mj12docs.htm.

14. U.S. National Advisory Committee for Aeronautics, *Forty First Annual Report of the National Advisory Committee for Aeronautics, 1955,* 283; JCH to Adm. John Callan, 31 January 1958, folder 42, box 1, JCHP/MIT; Jerome C. Hunsaker, *Some Lessons of History: Second Wings Club "Sight" Lecture* (New York: Wings Club, 1966), 22; JCH to Lee A. DuBridge, 31 March 1958, folder 4, box 24, JCHP/NASM.

15. Minutes of Regular Meeting of Executive Committee, 28 September 1956, pp. 5–6; Minutes of Regular Meeting of Executive Committee, 17 October 1956; both in Minutes, July–December 1956 folder, box 13, Minutes of the Executive Committee of the NACA, 1956–1958, Records of NASA, RG 255, NA.

16. Roland, *Model Research,* 1:290–96; JCH to Frederick C. Crawford, 23 April 1958, folder 42, box 1, JCHP/MIT; Minutes of Regular Meeting of Executive

Committee, 15 November 1956, p. 1, Minutes, July–December 1956 folder; Minutes of Regular Meeting of Executive Committee, 19 June 1958, p. 7, Minutes, January–June 1958 folder; Minutes of Regular Meeting of Executive Committee, 14 July 1958, Minutes, July–September 1958 folder; all in box 13, Minutes of the Executive Committee of the NACA, 1956–1958, Records of NASA, RG 255, NA.

17. Roland, *Model Research,* 1:209, 288, 302.
18. Stuart W. Leslie, *The Cold War and American Science: The Military-Industrial-Academic Complex at MIT and Stanford* (New York: Columbia University Press, 1993), 1–12; David F. Noble, *Forces of Production: A Social History of Industrial Automation* (New York: Oxford University Press, 1988), 3–20, 42–45.
19. For the motivation of people in the aircraft industry, see Jacob Vander Meulen, *The Politics of Aircraft: Building an American Military Industry* (Lawrence: University Press of Kansas, 1991), 45–46.
20. J. F. Victory to JCH, 26 October 1951; JCH speech on receipt of Wright Brothers Memorial Award, 17 December 1951; both in folder 35, box 1, JCHP/MIT.
21. JCH to Arthur A. Riley, 21 December 1953, folder 37, box 1, JCHP/MIT; NACA press release, 7 April 1955, Milton Ames Coll., file 71-4-2, Archives, NASA Langley Research Center, Hampton, Va. (courtesy of D. Bryan Taylor); Minutes of Regular meeting of Executive Committee, 17 January 1957, pp. 5–6, Minutes, January–June 1957 folder, box 13, Minutes of the Executive Committee of the NACA, 1956–1958, Records of NASA, RG 255, NA; *New York Times,* 25 May 1957.
22. JCH, "Aeronautical Engineering Department, MIT, 1933– 1960," pp. 13–14, folder 21, box 2, JCHP/MIT; Jerome C. Hunsaker, *Aeronautics at the Mid-Century* (New Haven: Yale University Press, 1952), 11–17, 24, 36–54, 98.
23. Hunsaker, *Aeronautics at the Mid-Century,* 74–80, 113–16.
24. See correspondence between JCH and Kranzberg, John B. Rae, and others, 1960–1967, folder 2, box 25, JCHP/NASM; *Technology and Culture* 1 (Winter 1959): 106–7; ibid., 3 (Summer 1962): 363–64; ibid., 4 (Winter 1963): 88–94. The 1962 review of the book on the NC-4 flight identified him as "James C. Hunsaker."
25. Lee M. Pearson to JCH, 19 December 1958; Pearson to JCH, 11 February 1960; JCH to Pearson, 9 February 1962; all in folder 7, box 3, JCHP/MIT.
26. Hunsaker, *Some Lessons of History,* 1–10.
27. Ibid., 15–16, 19, 24, 28–29.
28. J. P. Butterfield to JCH, 21 August 1958, folder 3, box 13, JCHP/NASM.
29. Minutes of First Meeting, Defense Group, Scientific Advisory Council, 10–11 November 1958; JCH to H. F. Sykes, 2 April 1959; both in folder 3, box 13, JCHP/NASM.
30. JCH memo to President's Office, MIT, 8 March 1963, folder 24, box 2; JCH, "Aeronautical Engineering Department, MIT, 1933–1960," pp. 3–4, folder 21, box 2; both in JCHP/MIT; JCH to Gilbert M. Roddy, 29 March 1963, folder 2, box 18; JCH to Carleton P. Fuller, 20 December 1968, folder 1, box 18; both in JCHP/NASM.
31. JCH to Frederick W. Barker, 29 November 1955, folder 39, box 1; JCH to George

Hewitt Myers, 30 June 1955, folder 39, box 1; JCH to Comdr. H. N. Slater, 3 January 1956, folder 40, box 1; JCH to Dr. J. T. Sample, 14 March 1958, folder 42, box 1; all in JCHP/MIT.

32. Robert C. Seamans Jr., interview by Walter Bonney, 23 February 1973, USAF Oral History Program, 866, USAF Historical Research Agency, Maxwell AFB, Ala.; J. W. Crowley to JCH, 4 May 1954, with itinerary of West Coast trip; JCH to Crowley, 20 March 1954; both in A-19, Personal File, Dr. Hunsaker, 1954, Correspondence and Report File, 1942–1960, box 207, Records of NASA, RG 255, NA; JCH to Rachel Carson, 30 November 1953, folder 12, box 10, JCHP/NASM; Bonney comments, 2 May 1972, on contents of box 8 of the Hunsaker Papers, NASM, J. C. Hunsaker file, Misc. Sources, NACA, NASA History Office.

33. *Technology Review* 62 (November 1959): 85, 112; JCH to H. K. Chow, 29 December 1959; JCH to Heinrich Peters, 29 December 1959; both in folder 43, box 1, JCHP/MIT.

34. JCH to J. M. Barker, 6 June 1968, folder 12, box 1; JCH to Donald Douglas, 21 October 1958, folder 42, box 1; JCH to Sir George Leigh-Jones, 15 February 1956, folder 40, box 1; all in JCHP/MIT.

35. JCH to Uncle Fred Clarke, 7 January 1958, folder 42, box 1; JCH to Mrs. Fred Easterday, 19 September 1958, folder 42, box 1; JCH to W. R. McKay, 20 October 1960, folder 23, box 2; all in JCHP/MIT; JCH to Col. G. W. Trichel, 7 December 1959, folder 3, box 13, JCHP/NASM; Bird, *Hunsaker Family History,* 16.

36. JCH to Mrs. Winfield Townsend, 24 February 1959, folder 43, box 1; Karl Arnstein to JCH, 28 March 1968, folder 12, box 3; both in JCHP/MIT; JCH to Dr. George W. Corner, 10 October 1966, folder 5, box 12, JCHP/NASM; Alice Hunsaker Bird to author, 10 June 1999. A brief obituary appears in the *New York Times,* 12 September 1966.

37. JCH to J. M. Barker, 6 June 1968, folder 12, box 1, JCHP/MIT.

38. JCH to Dr. George W. Corner, 11 January 1965, folder 4, box 12, JCHP/NASM; Jerome C. Hunsaker and Robert C. Seamans Jr., "Hugh Latimer Dryden (1898–1965)," in *American Philosophical Society Year Book 1966* (Philadelphia: American Philosophical Society, 1966), 121–27.

39. JCH to Adm. John Callan, 31 January 1958, folder 42, box 1, JCHP/MIT; JCH to Walter Bonney, 29 August 1972, J. C. Hunsaker file, box 2, Misc. Sources, NACA, NASA History Office; JCH to Eugene M. Emme, 6 March 1973, Hunsaker Biography File, NASA History Office; Alice Hunsaker Bird to author, 2 December 1996.

40. See copies of obituaries in Bird, *Jerome Clarke Hunsaker,* 179–80; Alice Hunsaker Bird to author, 10 June 1999; Robert C. Seamans Jr., "Jerome C. Hunsaker (1886–1984)," in *American Philosophical Society Year Book, 1985* (Philadelphia: American Philosophical Society, 1986), 132–37.

41. Hansen, "George W. Lewis," 98, 168.

BIBLIOGRAPHY

Manuscript and Archival Material

Hunsaker, Jerome C.
Biographical file. National Air and Space Museum Library, Washington, D.C.
Biographical file. Operational Archives Branch. Naval Historical Center, Washington, D.C.
Biography File. History Office. National Aeronautics and Space Administration Headquarters, Washington, D.C.
Papers. MC 272. Institute Archives and Special Collections. Massachusetts Institute of Technology, Cambridge, Mass.
Papers. National Air and Space Museum Archives, Washington, D.C.

National Archives and Records Administration
Bureau of Aeronautics. General Correspondence, 1925–1942, Record Group 72.
———. General Correspondence initiated in the Bureau of Construction and Repair, 1917–1925, Record Group 72.
———. Office Services Division. Records of the Bureau of Aeronautics, Record Group 72.
Deck Logs (*Arkansas, Cleveland, Hartford*), Record Group 24.
Hunsaker Biography File. Records of the National Aeronautics and Space Administration, Record Group 255.
Intelligence Division. Naval Attaché Reports, 1886–1939. Records of the Deputy Chief of Naval Operations, 1882–1954, Record Group 38.
Military Intelligence Division Correspondence, 1917–1941. Records of the War Department General and Special Staffs. Record Group 165.

Minutes of the Executive Committee of the NACA, 1943–1948, 1956–1958. Records of the National Aeronautics and Space Administration, Record Group 255.
National Advisory Committee for Aeronautics. General Correspondence Files. Records of the National Aeronautics and Space Administration, Record Group 255.
Office of Naval Intelligence. Records, Record Group 38.
Personal File, Dr. Hunsaker. Correspondence and Report File, 1942–1960. Records of the National Aeronautics and Space Administration, Record Group 255.
Proceedings of Naval and Marine Examining Boards. Records of the Office of the Judge Advocate General (Navy), Record Group 125.
Records Relating to NACA Committees and Subcommittees (Decimal File), 1918–1951. Records of the National Aeronautics and Space Administration, Record Group 255.
Secretary of the Navy. General Correspondence, Record Group 80.

Other
Ames, Milton. Collection. File 71-4-2. Archives. National Aeronautics and Space Administration. Langley Research Center, Hampton, Va.
Furer, Julius A. Papers. Library of Congress. Manuscript Division.
General Board. Records. File 449. Operational Archives Branch. Naval Historical Center, Washington, D.C. (now located at the National Archives).
Guggenheim Records. Library of Congress. Manuscript Division.
Hearings before the General Board of the Navy, 1919–1922 (microfilm rolls 2, 3, 5).
Lighter-than-Air Society. Collection. University of Akron Archives, Akron, Ohio.
Mead, George J. Miscellaneous Sources. NASA History Office. NASA Headquarters, Washington, D.C.
Millikan, Clark B. Papers. Institute Archives. California Institute of Technology, Pasadena, Calif.
Moffett, William A. Papers. Ms. Coll. 198 (microfilm). Nimitz Library. U.S. Naval Academy, Annapolis, Md.
Record of Officers, U.S. Navy. File 6733. National Personnel Records Center, St. Louis, Mo.
Records of Midshipman Jerome Clarke Hunsaker, Record Group 405. William W. Jeffries Memorial Archives. U.S. Naval Academy, Annapolis, Md.
WPD 888-02. File 145.93-97. U.S. Air Force Historical Research Agency. Maxwell Air Force Base, Alabama.

Public Documents
Annual Register of the United States Naval Academy, 1905–1906, 1906–1907, 1907–1908, 1908–1909. Washington, D.C.: Government Printing Office, 1905, 1906, 1907, 1908.
Hunsaker, Jerome C. *Dynamical Stability of Aeroplanes*. Smithsonian Miscellaneous Collections, vol. 62, no. 5. Washington, D.C.: Smithsonian Institution, 1916.

United States

The Airport and Its Neighbors: The Report of the President's Airport Commission. Washington, D.C.: Government Printing Office, 1952.

Federal Aviation Commission. *Report of the Federal Aviation Commission.* Washington, D.C.: Government Printing Office, 1935.

National Advisory Committee for Aeronautics. *Twenty-Seventh Annual Report of the National Advisory Committee for Aeronautics, 1941.* Washington, D.C.: Government Printing Office, 1942.

————. *Twenty-Eighth Annual Report of the National Advisory Committee for Aeronautics, 1942.* Washington, D.C.: Government Printing Office, 1946.

————. *Fortieth Annual Report of the National Advisory Committee for Aeronautics, 1954.* Washington, D.C.: Government Printing Office, 1956.

————. *Forty-First Annual Report of the National Advisory Committee for Aeronautics, 1955.* Washington, D.C.: Government Printing Office, 1956.

President's Aircraft Board. *Hearings before the President's Aircraft Board.* 4 vols. Washington, D.C.: Government Printing Office, 1925.

Survival in the Air Age: A Report by the President's Air Policy Commission. Washington, D.C.: Government Printing Office, 1948.

U.S. Congress, House

Congressional Record. Vol. 75, pts. 11, 12. Washington, D.C.: Government Printing Office, 1932.

Hearings before the Committee on Naval Affairs . . . on Estimates Submitted by the Secretary of the Navy, 1916. 2 vols. Washington, D.C.: Government Printing Office, 1916.

Hearings before the Committee on Naval Affairs . . . on Estimates Submitted by the Secretary of the Navy, 1920. 2 vols. Washington, D.C.: Government Printing Office, 1920.

Hearings before the Committee on Naval Affairs . . . on Sundry Legislation Affecting the Naval Establishment, 1920–1921. Washington, D.C.: Government Printing Office, 1921.

Hearings before the Select Committee of Inquiry into Operations of the United States Air Services. Washington, D.C.: Government Printing Office, 1925.

Hearings before the Subcommittee of the Committee on Appropriations on Second Supplemental National Defense Appropriations Bill for 1942. Washington, D.C.: Government Printing Office, 1941.

Hearings before the Select Committee on Post-War Military Policy. Washington, D.C.: Government Printing Office, 1945.

Hearings before the Subcommittee of the Committee on Appropriations on the Independent Offices Appropriation Bill for 1949. Washington, D.C.: Government Printing Office, 1948.

Hearings before the Subcommittee of the Committee on Appropriations on the Inde-

pendent Offices Appropriation Bill for 1950. Washington, D.C.: Government Printing Office, 1949.

Hearings before the Subcommittee of the Committee on Appropriations on the Independent Offices Appropriation Bill for 1951. Washington, D.C.: Government Printing Office, 1950.

Hearings before the Subcommittee of the Committee on Appropriations on the Independent Offices Appropriation Bill for 1952. Washington, D.C.: Government Printing Office, 1951.

Hearings before the Subcommittee of the Committee on Appropriations on the Independent Offices Appropriation Bill for 1953. Washington, D.C.: Government Printing Office, 1952.

Merchants' Aircraft. Hearing before the Committee on Interstate and Foreign Commerce. Washington, D.C.: Government Printing Office, 1931.

United Air Service. Hearing before a Subcommittee of the Committee on Military Affairs. Washington, D.C.: Government Printing Office, 1919.

U.S. Congress, Senate

Congressional Record. Vol. 72, pt. 7. Washington, D.C.: Government Printing Office, 1930.

Congressional Record. Vol. 76, pt. 5. Washington, D.C.: Government Printing Office, 1933.

Hearings before a Joint Committee to Investigate Dirigible Disasters. Washington, D.C.: Government Printing Office, 1933.

Hearings before the Subcommittee of the Committee on Appropriations, Third Supplemental National Defense Appropriation Bill for 1942. Washington, D.C.: Government Printing Office, 1941.

Hearings before a Special Committee Investigating the National Defense Program. Washington, D.C.: Government Printing Office, 1946.

Merchants' Aircraft. Hearing before the Committee on Commerce. Washington, D.C.: Government Printing Office, 1931.

Newspapers and Periodicals

Aero Digest
Aeronautical Journal
Aeronautics
Aircraft
Aviation
Journal of the Aeronautical Sciences
New York Times
Opelika (Ala.) *Daily News*
Tech (MIT)

Technology and Culture
Technology Review
Times (London)
Weekly News Letter (Bureau of Aeronautics)

Books

The Aircraft Year Book for 1931. New York: Aeronautical Chamber of Commerce of America, 1931.

Aitken, Hugh G. J. *Scientific Management in Action: Taylorism at Watertown Arsenal*. Princeton: Princeton University Press, 1985.

Anderson, John D., Jr. *A History of Aerodynamics and Its Impact on Flying Machines*. Cambridge: Cambridge University Press, 1997.

Barlow, Jeffrey G. *Revolt of the Admirals: The Fight for Naval Aviation, 1945–1950*. Washington, D.C.: Naval Historical Center, 1994.

Bates, Charles C., and John F. Fuller. *America's Weather Warriors, 1814–1985*. College Station: Texas A&M University Press, 1986.

Bird, Alice Hunsaker. *Jerome Clarke Hunsaker: A Life in Aeronautics*. Henley-on-Thames, England: Privately published, 1982.

———. *Hunsaker Family History*. Henley-on-Thames, England: Privately published, 1995.

Boyce, Joseph C., ed. *New Weapons for Air Warfare: Fire-Control Equipment, Proximity Fuzes, and Guided Missiles*, in *Science in World War II, Office of Scientific Research and Development*. Boston: Little, Brown, 1947.

Brooks, Peter W. *The Modern Airliner: Its Origins and Development*. London: Putnam, 1961.

Caute, David. *The Great Fear: The Anti-Communist Purge under Truman and Eisenhower*. New York: Simon and Schuster, 1978.

Coletta, Paolo E. *Admiral Bradley A. Fiske and the American Navy*. Lawrence: Regents Press of Kansas, 1979.

Constant, Edward W., II. *The Origins of the Turbojet Revolution*. Baltimore: Johns Hopkins University Press, 1980.

Cunningham, Frank. *Skymaster: The Story of Donald Douglas*. Philadelphia: Dorrance, 1943.

Current Biography, 1942. New York: H. W. Wilson, 1942.

Davies, R. E. G. *Airlines of the United States since 1914*. London: Putnam, 1972.

Davis, Burke. *The Billy Mitchell Affair*. New York: Random House, 1967.

Davis, George T. *A Navy Second to None: The Development of Modern American Naval Policy*. New York: Harcourt, Brace, 1940.

Dawson, Virginia P. *Engines and Innovation: Lewis Laboratory and American Propulsion Technology*. Washington, D.C.: National Aeronautics and Space Administration, 1991.

Dick, Harold G., and Douglas H. Robinson. *The Golden Age of the Great Passenger Airships: Graf Zeppelin and Hindenburg.* Washington, D.C.: Smithsonian Institution Press, 1985.

Dictionary of American Naval Fighting Ships. Vol. 6. Washington, D.C.: Naval History Division, 1976.

Dyer, George Carroll. *The Amphibians Came to Conquer: The Story of Admiral Richmond Kelly Turner.* 2 vols. Washington, D.C.: Department of the Navy, 1972.

Foxworth, Thomas G. *The Speed Seekers.* New York: Doubleday, 1976.

Francillon, Rene J. *McDonnell Douglas Aircraft since 1920.* Annapolis: Naval Institute Press, 1988.

Furer, Julius Augustus. *Administration of the Navy Department in World War II.* Washington, D.C.: Naval History Division, U.S. Navy Department, 1959.

Gorn, Michael H. *The Universal Man: Theodore von Kármán's Life in Aeronautics.* Washington, D.C.: Smithsonian Institution Press, 1992.

Graham, Margaret, and Bettye H. Pruitt. *R&D for Industry: A Century of Technical Innovation at Alcoa.* Cambridge: Cambridge University Press, 1990.

Gross, Stuart D. *Saginaw: A History of the Land and the City.* Woodland Hills, Calif.: Windsor, 1980.

Hallion, Richard P. *Legacy of Flight: The Guggenheim Contribution to American Aviation.* Seattle: University of Washington Press, 1977.

———. *On the Frontier: Flight Research at Dryden, 1946–1981.* Washington, D.C.: National Aeronautics and Space Administration, 1984.

Hanle, Paul A. *Bringing Aerodynamics to America.* Cambridge: MIT Press, 1982.

Hansen, James R. *Engineer in Charge: A History of the Langley Aeronautical Laboratory, 1917–1958.* Washington, D.C.: National Aeronautics and Space Administration, 1987.

Hansen, James R. "George W. Lewis and the Management of Aeronautical Research." In William M. Leary, ed., *Aviation's Golden Age: Portraits from the 1920s and 1930s.* Iowa City: University of Iowa Press, 1989.

Holland, Maurice. *Architects of Aviation.* New York: Duell, Sloan, and Pearce, 1951.

Holley, Irving Brinton, Jr. *Ideas and Weapons: Exploitation of the Aerial Weapon by the United States during World War I.* New Haven: Yale University Press, 1953.

Hunsaker, Jerome C. "Experimental Analysis of Inherent Longitudinal Stability for a Typical Biplane." In *First Annual Report of the National Advisory Committee for Aeronautics, 1915,* 25–51. Washington, D.C.: National Advisory Committee for Aeronautics, 1916.

———. *Aeronautics at the Mid-Century.* New Haven: Yale University Press, 1952.

———. *Forty Years of Aeronautical Research.* From *Smithsonian Report for 1955.* Washington, D.C.: Smithsonian Institution, 1956.

———. *Some Lessons of History: Second Wings Club "Sight" Lecture.* New York: Wings Club, 1966.

Hunsaker, J. C., and B. G. Rightmire. *Engineering Applications of Fluid Mechanics.* New York: McGraw-Hill, 1947.

Jakab, Peter L. *Visions of a Flying Machine: The Wright Brothers and the Process of Invention*. Washington, D.C.: Smithsonian Institution Press, 1990.

Kanigel, Robert. *The One Best Way: Frederick Winslow Taylor and the Enigma of Efficiency*. New York: Viking, 1997.

Kilar, Jeremy W. *Michigan's Lumbertowns: Lumbermen and Laborers in Saginaw, Bay City, and Muskegon, 1870–1905*. Detroit: Wayne State University Press, 1990.

Komons, Nick A. *Bonfires to Beacons: Federal Civil Aviation Policy under the Air Commerce Act, 1926–1938*. Washington, D.C.: Federal Aviation Administration, Department of Transportation, 1978.

Layton, Edwin T., Jr. *The Revolt of the Engineers: Social Responsibility and the American Engineering Profession*. Cleveland: Press of Case Western Reserve University, 1971.

Leslie, Stuart W. *The Cold War and American Science: The Military-Industrial-Academic Complex at MIT and Stanford*. New York: Columbia University Press, 1993.

Loyrette, Henri. *Gustave Eiffel*. Translated by Rachel Gomme and Susan Gomme. New York: Rizzoli International, 1985.

Lucky Bag. Vol. 13. Springfield, Mass.: F. A. Bassette, 1906.

————. Vol. 15. Philadelphia: Hoskins Press, 1908.

MacKenzie, Donald. *Inventing Accuracy: A Historical Sociology of Nuclear Missile Guidance*. Cambridge: MIT Press, 1990.

Melhorn, Charles M. *Two-Block Fox: The Rise of the Aircraft Carrier, 1911–1929*. Annapolis: Naval Institute Press, 1974.

Miller, Jay. *Lockheed U-2*. Austin, Tex.: Aerofax, 1983.

Mills, James Cooke. *History of Saginaw County, Michigan: Historical, Commercial, Biographical*. 2 vols. Saginaw, Mich.: Seeman and Peters, 1918.

Montague, Ludwell Lee. *General Walter Bedell Smith as Director of Central Intelligence*. University Park: Pennsylvania State University Press, 1992.

Morrison, Wilbur H. *Donald W. Douglas: A Heart with Wings*. Ames: Iowa State University Press, 1991.

Muenger, Elizabeth A. *Searching the Horizon: A History of Ames Research Center, 1940–1976*. Washington, D.C.: National Aeronautics and Space Administration, 1985.

Munson, Kenneth, and Gordon Swanborough. *Boeing: An Aircraft Album*. No. 4. New York: Arco, 1972.

The National Cyclopedia of American Biography. Vol. 43. New York: James T. White, 1961.

Noble, David F. *America by Design: Science, Technology, and the Rise of Corporate Capitalism*. New York: Knopf, 1977.

————. *Forces of Production: A Social History of Industrial Automation*. New York: Oxford University Press, 1988.

Packard, Wyman H. *A Century of U.S. Naval Intelligence*. Washington, D.C.: Office of Naval Intelligence and Naval Historical Center, 1996.

Pearson, Henry Greenleaf. *Richard Cockburn Maclaurin: President of the Massachusetts Institute of Technology.* New York: Macmillan, 1937.

Prescott, Samuel C. *When MIT Was "Boston Tech," 1861–1916.* Cambridge, Mass.: Technology Press, 1954.

Register of Alumni, 1845–1956: Graduates and Former Naval Cadets and Midshipmen. Annapolis: U.S. Naval Academy Alumni Association, 1956.

Reich, Leonard S. *The Making of American Industrial Research: Science and Business at GE and Bell, 1876–1926.* Cambridge: Cambridge University Press, 1985.

Reynolds, Clark G. *Admiral John H. Towers: The Struggle for Naval Air Supremacy.* Annapolis: Naval Institute Press, 1991.

Rich, Ben R., and Leo Janos. *Skunk Works: A Personal Memoir of My Years at Lockheed.* Boston: Little, Brown, 1994.

Robinson, Douglas H. *Giants in the Sky: A History of the Rigid Airship.* Seattle: University of Washington Press, 1973.

Robinson, Douglas H., and Charles L. Keller. *"Up Ship!" U.S. Navy Rigid Airships, 1919–1935.* Annapolis: Naval Institute Press, 1982.

Roland, Alex. *Model Research: The National Advisory Committee for Aeronautics, 1915–1958.* 2 vols. Washington, D.C.: National Aeronautics and Space Administration, 1985.

Roseberry, C. R. *Glenn Curtiss: Pioneer of Flight.* Garden City, N.Y.: Doubleday, 1972.

Schatzberg, Eric. *Wings of Wood, Wings of Metal: Culture and Technical Choice in American Airplane Materials, 1914–1945.* Princeton: Princeton University Press, 1999.

Serling, Robert J. *The Only Way to Fly: The Story of Western Airlines, America's Senior Air Carrier.* Garden City, N.Y.: Doubleday, 1976.

Smith, Richard K. *The Airships* Akron *and* Macon: *Flying Aircraft Carriers of the United States Navy.* Annapolis: Naval Institute Press, 1965.

———. *First Across! The U.S. Navy's Transatlantic Flight of 1919.* Annapolis: Naval Institute Press, 1973.

Snyder, Bernard J. *Aircraft Nuclear Propulsion: An Annotated Bibliography.* Washington, D.C.: USAF History and Museums Program, 1996.

Sweetman, Jack. *The U.S. Naval Academy: An Illustrated History.* Annapolis: Naval Institute Press, 1979.

Trimble, William F. *Wings for the Navy: A History of the Naval Aircraft Factory, 1917–1956.* Annapolis: Naval Institute Press, 1990.

———. *Admiral William A. Moffett: Architect of Naval Aviation.* Washington, D.C.: Smithsonian Institution Press, 1994.

Turnbull, Archibald D., and Clifford L. Lord. *History of United States Naval Aviation.* New Haven: Yale University Press, 1949.

Van der Linden, F. Robert. "Progressives and the Post Office: Walter Folger Brown and the Creation of United States Air Transportation." In William F. Trimble, ed.,

From Airships to Airbus: The History of Civil and Commercial Aviation. Vol. 2, *Pioneers and Operations.* Washington, D.C.: Smithsonian Institution Press, 1995.

Vander Meulen, Jacob. *The Politics of Aircraft: Building an American Military Industry.* Lawrence: University Press of Kansas, 1991.

Vincenti, Walter G. *What Engineers Know and How They Know It: Analytical Studies from Aeronautical History.* Baltimore: Johns Hopkins University Press, 1990.

Von Kármán, Theodore, with Lee Edson. *The Wind and Beyond: Theodore von Kármán, Pioneer in Aviation and Pathfinder in Space.* Boston: Little, Brown, 1967.

Westervelt, G. C., H. C. Richardson, and A. C. Read. *The Triumph of the N.Cs.* Garden City, N.Y.: Doubleday, Page, 1920.

Whisler, Timothy R. *The British Motor Industry, 1945–1994: A Case Study in Industrial Decline.* Oxford: Oxford University Press, 1999.

Whitnah, Donald R. *A History of the United States Weather Bureau.* Urbana: University of Illinois Press, 1961.

Wilson, Eugene E. *Slipstream: The Autobiography of an Air Craftsman.* 3d ed. Palm Beach, Fla.: Literary Investment Guild, 1967.

Wilson, John R. M. *Turbulence Aloft: The Civil Aeronautics Administration Amid Wars and Rumors of Wars, 1938–1953.* Washington, D.C.: Federal Aviation Administration, U.S. Department of Transportation, 1979.

Zachary, G. Pascal. *Endless Frontier: Vannevar Bush, Engineer of the American Century.* Cambridge: MIT Press, 1999.

Zahm, A. F. *Report on European Aeronautical Laboratories.* Smithsonian Miscellaneous Collections, vol. 62, no. 3. Washington, D.C.: Smithsonian Institution, 1914.

Articles

Armstrong, William J. "Dick Richardson: His Life in Aeronautics." *Naval Aviation News* (April 1977): 34–37.

Byers, Horace R. "Carl-Gustaf Arvid Rossby." In *Biographical Memoirs.* Vol. 34. New York: National Academy of Sciences, 1960.

Carlson, W. Bernard. "Academic Entrepreneurship and Engineering Education: Dugald C. Jackson and MIT-GE Cooperative Engineering Course, 1907–1932." *Technology and Culture* 29 (July 1988): 536–67.

Daniels, R. W. "Duralumin." *Society of Automotive Engineers, Transactions* 17, pt. 2 (1922): 751–58.

Dickey, Fred C., Jr. "U.S. Navy Ship Plane Units." *Journal of the American Aviation Historical Society* 10 (Summer 1965): 121–26.

D'Orcy, Ladislas. "The Curtiss Marine Flying Trophy Race." *Aviation* 13 (16 October 1922): 490–92.

Fiske, Bradley A. "Navigating without Horizon." *United States Naval Institute Proceedings* 33 (September 1907): 955–57.

Hanson, R. T., and J. C. Hunsaker. "Rudder Trials, USS *Sterett.*" *Transactions: Society of Naval Architects and Marine Engineers* 20 (New York: Society of Naval Architects and Marine Engineers) (1912): 301–34.

Heyser, William. "Technology's Wind Tunnels." *Tech Engineering News* 21 (January 1941): 341, 349.

Hunsaker, J. C. "Europe's Facilities for Aeronautical Research—I." *Flying* 3 (April 1914): 75, 93.

———. "Europe's Facilities for Aeronautical Research—II." *Flying* 3 (May 1914): 108–9.

———. "The Present Status of Air-Ships in Europe." *Journal of the Franklin Institute* 177 (June 1914): 597–639.

———. "Scientific Aeronautic Research: The New Aerodynamic Laboratory of the Massachusetts Institute of Technology." *Scientific American Supplement,* no. 2057 (5 June 1915): 364–65.

———. "Notes on Aeroplane Shock Absorbers of Rubber." *Aeronautics* 11 (London) (11 October 1916): 235–39.

———. "Progress in Naval Aircraft." *Journal of the Society of Automotive Engineers* 5 (July 1919): 31–44.

———. "The Navy's First Airships." *United States Naval Institute Proceedings* 45 (August 1919): 1347–68.

———. "Progress in Naval Aircraft." *Society of Automotive Engineers, Transactions* 14, pt. 2 (1919): 236–77.

———. "Naval Architecture in Aeronautics." *Aeronautical Journal* 24 (July 1920): 321–405.

Hunsaker, Jerome C., and George J. Mead. "Around the Corner in Aviation." *Technology Review* 39 (December 1936): 65–68, 80.

Hunsaker, Jerome C., and Robert C. Seamans Jr. "Hugh Latimer Dryden (1898–1965)." In *American Philosophical Society Year Book, 1966.* Philadelphia: American Philosophical Society, 1966.

Klass, Philip J. "Special Report: The MJ-12 Crashed—Saucer Documents." *Skeptical Inquirer* 12 (Winter 1987–88): 137–46.

Lapointe, E. "Aviation in the Navy." *United States Naval Institute Proceedings* 38 (June 1912): 627–56.

Martin, D. K. "Laying a Foundation for Aircraft Communication." *Bell Laboratories Record* 7 (April 1929): 315–18.

Moffett, William A. "Organization and Function of Naval Aviation." *Aviation* 13 (28 August 1922): 248–49.

Post, Robert C. Interview with Walter Vincenti. "What Engineers Know." *Invention and Technology* 12 (Winter 1997): 20–26.

Rose, Don. "Bigger and Better Balloons." *Aero Digest* 14 (June 1929): 57–58, 252–56.

Rossby, Carl Gustaf. "Airways and the Weather." *Western Flying* 4 (April 1928): 23–24, 96.

Schatzberg, Eric. "Ideology and Technical Choice: The Decline of the Wooden Airplane in the United States, 1920–1945." *Technology and Culture* 35 (January 1994): 34–69.

Seamans, Robert C., Jr. "Jerome C. Hunsaker (1886–1984)." In *American Philosophical Society Year Book, 1985*. Philadelphia: American Philosophical Society, 1986.

Smith, Richard K. "A Douglas Decision." *Aerospace Historian* 15 (Autumn 1968): 5–7.

———. "The Intercontinental Airliner and the Essence of Airplane Performance, 1929–1939." *Technology and Culture* 24 (July 1983): 428–49.

———. "The Superiority of the American Transoceanic Airliner, 1932–1939: Sikorsky S-42 versus Short S. 23." *American Aviation Historical Society Journal* 29 (Summer 1984): 82–94.

———. "The Weight Envelope: An Airplane's Fourth Dimension . . . Aviation's Bottom Line." *Aerospace Historian* 33 (Spring 1986): 30–44.

———. "Notes on the International Zeppelin Transport Co. (IZT) and Pacific Zeppelin Transport Co. (PZT)." *Buoyant Flight* 39 (July–August 1992): 4–5.

Thornton, Earl A. "MIT, Jerome C. Hunsaker, and the Origins of Aeronautical Engineering." *American Aviation Historical Society Journal* 43 (Winter 1998): 306–13.

Trimble, William F. "Jerome C. Hunsaker, Bell Labs, and the West Coast Model Airline." *Journal of the West* 36 (July 1997): 44–52.

Vincenti, Walter G. "The Retractable Landing Gear and the Northrop 'Anomaly': Variation-Selection and the Shaping of Technology." *Technology and Culture* 35 (January 1994): 1–33.

"A Visit to Friedrichshafen." *Gasbag Journal*, no. 37 (September 1998): 23–26.

Interviews and Oral Histories

Barnaby, Capt. Ralph S. (Ret.). Interview by the author, 12 October 1983, Philadelphia.

Hunsaker, Jerome C. Interview by Walter T. Bonney, 2 November 1971. Hunsaker, Jerome C. Biography File. NASA History Office. NASA Headquarters, Washington, D.C.

Hunsaker, Jerome C., Jr. Interview by Earl A. Thornton, 12 November 1996, Lincoln, Mass.

"The Reminiscences of Jerome C. Hunsaker." Interview by Donald Shaughnessy, April 1960. Aviation Project, Oral History Research Office, Columbia University.

Seamans, Robert C., Jr. Interview by Walter Bonney, 23 February 1973. USAF Oral History Program, 866. USAF Historical Research Agency. Maxwell AFB, Ala.

Dissertations, Papers, Web Sites, and Other Unpublished Materials

Dong, Jingsheng. "Biography of Hsue-shen Tsien: The Story of an American-Trained Chinese Rocket Expert." Ph.D. diss., Auburn University, 1996.

Fahrney, D. S. "History of Pilotless Aircraft and Guided Missiles." Undated ms., probably 1958. Operational Archives Branch. Naval Historical Center, Washington, D.C.

Hansen, James R. "The Revolt against Max Munk." Undated manuscript. Copy in files of James R. Hansen. Auburn University, Auburn, Ala.

Levinson, Mark. "Encounters between Engineering Science and American Engineering Education: The Postwar Textbook Revolution." Paper presented to the Society for the History of Technology, Pasadena, Calif., 16–19 October 1997.

Lord, Clifford L. "The History of Naval Aviation, 1898–1939." 1946. Office of the Deputy Chief of Naval Operations (Air). Naval Aviation History Unit, Washington, D.C.

Majestic 12. http://site034145.primehost.com/ds/mj12docs.htm.

Ober, Shatswell. "The Story of Aeronautics at MIT, 1895 to 1960." 1965. Department of Aeronautics and Astronautics. MIT, Cambridge, Mass.

———. "Development of Aeronautical Engineering at MIT." Period IV. May 1965. Supplement to "The Story of Aeronautics at MIT." Department of Aeronautics and Astronautics. MIT, Cambridge, Mass.

Shanahan, William O. "Procurement of Naval Aircraft, 1907–1939." Vol. 17. 1946. Deputy Chief of Naval Operations (Air). Naval Aviation History Unit, Washington, D.C.

Smith, Richard K. "Zeppelin Kaput!" About 1988. In possession of Richard K. Smith.

"Union County History." Union County (Iowa) Genealogical Society. http://lserver. aea14.k12.ia.us/SWP/jbriley/ucgen/history.html.

World War II Administrative History, BuAer. Vol. 1, Background. 1957. Department of the Navy, Washington, D.C.

INDEX

Aero Club of America, 85
aerodynamics, 24–37, 56, 149–52, 156; compressibility, 149, 158, 171, 174–75, 184; high-speed, 209; hypersonic, 192, 193, 194; laminar-flow wing, 188; supersonic, 181, 187–88, 189, 199
Aerodynamics Institute, Aachen, 120
Aeromarine, 50
aeronautics, as "high tech," 2, 22
Aeronautics Branch, Department of Commerce, 104, 105, 107, 109, 141
Ahlborn, Friedrich, 29
Air Commerce Act, 104
aircraft: engines, 171–75, 188; nuclear propulsion, 196–97; procurement, 48–50
Aircraft Division, Bureau of Construction and Repair, 37–38, 39, 48, 49, 50, 60, 63, 76; expansion and reorganizaton, 52–53; NC flying boats, 54; rigid airship, 66
aircraft carrier, 58, 62
aircraft industry, 50, 178–79, 183, 184–85
Aircraft Production Board, 48–49
Air Defense League, 141
Air Engineering Development Center (AAF), 189–90, 192, 194, 195
air force, 194, 195–96
airlines, 108–10, 119, 128, 129, 142–43, 149, 151, 154–57, 165; and passengers, 100, 101–2, 105, 106; safety, 197–99
airmail controversy (1934), 140–41
airplanes, types: Albatross, 155; Bell XS-1, 191; Boeing 247, 150; Boeing 300, 150–51; Boeing 307 Stratoliner, 151, 156; 151; Boeing 314, 155, 156; B-36, 195; Consolidated PB2Y-2, 155; Curtiss CR-2, 86–87; Curtiss CR-3, 87–88; Curtiss F-5-L, 49–50; Curtiss HS-1, 49, 50; Curtiss H-16, 49, 50; Curtiss JN-2, 34, 36; Curtiss JN-4B, 40; Curtiss N-9, 40–41; Curtiss R-6, 49, 87; Curtiss R2C-1, 88; De Havilland DH.108, 188; Dornier Rs.IV, 61; Douglas C-54, 157; Douglas DC-4E, 150–51; Douglas DC series, 150; Douglas D-558-1, 191; Douglas DT, 73; Fokker F-10, 106; Gloster E.28/39, 172; Lockheed L-44 Excalibur, 155–56; Lockheed U-2, 211; Martin 130, 155; Martin 162, 155; NC flying boats, 53–56; North American P-51, 183; NW "mystery" racer, 87; NW-2, 87–88; Sikorsky PBS-1, 155, 156; Sikorsky S-42, 149; Sikorsky S-42A, 149; Sikorsky S-42B, 152; Sikorsky XPBS-1, 154; Speed Scout, 41; TR-1, 86, 87; TR-3, 86; TS-1, 62–63, 86; TS-2, 86; Vought-Sikorsky VS-44, 156; Wright F2W-1, 88
airports, 198–99, 216
air power and strategic bombing, 195–96, 215
air racing, 73, 85–88
airship: acquisition from British, 63–64, 65–66; A-type, 44; B-type, 44–45, 51; commercial, 112–19, 122–31, 133, 134–35, 142, 144, 154; congressional investigation, 132–34, 140; C-type, 51–52; as dead-end technology, 154–55; DN-1, 43; federal subsidy of, 124–25; Fleet Airship 1, 66, 72; German, 28–29, 81–82, 127; GZ-3, 131; L.2, 29; L.49, 61, 66; L.70, 66; L.72, 66; LZ.128, 122–24; military, 143; nonrigid, 42–45, 50–52;